KS3
Success

Maths

Complete Coursebook

Fiona Mapp
Pamela Wild

Age 11-14

Contents

Contents

Chapter 7 Measurement and its applications

Chapter 8 Probability

Chapter 9 Data collection

Chapter 10 Representing and interpreting data

1 Types of numbers and their applications

After studying this chapter you should be able to:
- recognise and describe number relationships including multiple, factor and square
- use and apply fractions and decimals in a variety of problems
- use and apply percentages in everyday problems
- solve problems involving ratio and proportion
- solve problems that involve calculating with powers, roots and numbers expressed in standard form.

 1.1 # Numbers, powers and roots

Place value

Each digit in a number has a place value. The value of the digit depends on its place in the number.

The place value changes by a factor of 10 as you move from one digit to the next.

538	2371	6 352 740
Five hundred and thirty-eight	Two thousand, three hundred and seventy-one	Six million three hundred and fifty-two thousand seven hundred and forty
The digit 5 represents five hundred	The digit 7 represents seven tens or seventy	The digit 5 represents fifty thousand

Here are some tips for ordering whole numbers:
1. Put the numbers into groups with the same number of digits.
2. For each group, arrange each number in order of size depending on the place value of the digits.

Example

Arrange these numbers in order of size, smallest first:

26, 502, 794, 3297, 4209, 4351, 7 459 080, 5, 32, 85, 114, 54 321

This becomes:

5, 26, 32, 85, 114, 502, 794, 3297, 4209, 4351, 54 321, 7 459 080

Practise arranging numbers in order of size. Make four sets of cards, each numbered with the digits 0–9. Shuffle them. Put out four cards to make a four-digit number. Say the number out loud. Put out three more four-digit numbers, reading each one out loud. Now order them from highest to lowest. Repeat the exercise with three, five and six-digit numbers. Alternate ordering highest to lowest with lowest to highest.

Binary numbers

Binary is the language used by computers. It uses 0 and 1 to represent different numbers.

In the everyday number system, we use 0 – 9 to show numbers.

H	T	U
5	6	9

If a number has two whole numbers to its right, we know that it has a 'hundreds' value. For example, the 5 above represents 500. This number system is known as 'Base 10'. Each position to the left is worth 10× more than the place to the right:

Binary uses 'Base 2'. Each position to the left is worth 2× more than the place to the right:

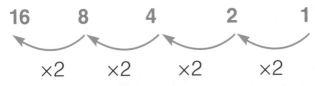

Of course you don't have to write in 1, 2, 4, etc. You just remember them like you remember HTU.

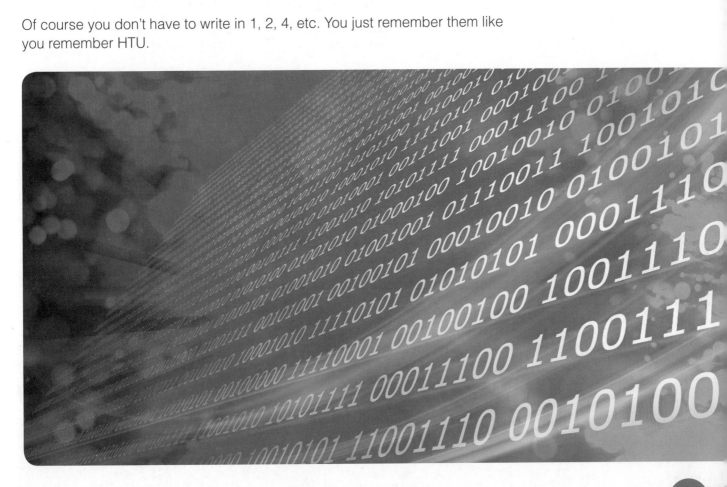

Example

Write 9 as a binary number.

16	8	4	2	1
	1	0	0	1

You have to see which columns you can use to get to 9.
Put a 1 in the place that has the value of '8', zeros in the '4' and '2' columns and a 1 in the '1' column. (1×8) and (1×1) gives 9.

Here are some other examples of binary numbers:

32	16	8	4	2	1		
		1	0	1	0	=	10
	1	0	0	1	1	=	19
1	0	0	0	0	1	=	33

Directed numbers

Positive and negative numbers are used on the coordinate axes. See Topic 4.2.

Integers are whole numbers that can be positive or negative. Positive numbers are above zero. Negative numbers are below zero. Integers are sometimes known as **directed numbers**.

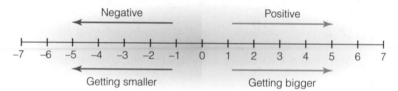

For example:

−10	is smaller than	−8	so	−10	<	−8
−4	is bigger than	−6	so	−4	>	−6
5	is bigger than	−2	so	5	>	−2

Directed numbers are often seen on the weather forecast in winter.

On this weather map, Aberdeen is the coldest place at −8°C and London is 6 degrees warmer than Manchester.

Adding and subtracting directed numbers

Look at the following:
The temperature at 6am was −5°C.
By 10am it had risen 8 degrees.
$-5° + 8° = 3°$
So the new temperature was 3°C.

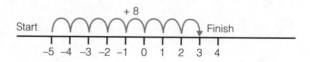

It is useful to draw a number line to help when answering questions of this type.

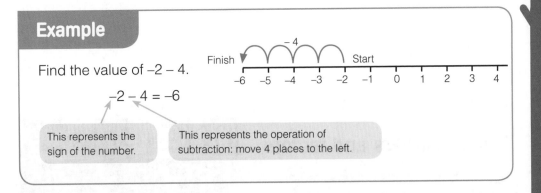

Example

Find the value of −2 − 4.

−2 − 4 = −6

This represents the sign of the number.

This represents the operation of subtraction: move 4 places to the left.

Practise moving along a number line with + and − numbers. Make cards from −10 to +10. Place them vertically on the floor. Ask a friend to tell you to stand by any number and to move + or − a number of sensible spaces. Say out loud what the move entailed and where you ended up. For example: Stand by −6. Move +3. This is −6 + 3 = −3

When the number to be added (or subtracted) is negative, the normal direction of movement is reversed.

When adding and subtracting directed numbers:

Like signs give an addition	Unlike signs give a subtraction
+ (+) = +	− (+) = −
− (−) = +	+ (−) = −

For example:

−2 + (−3) = −2 − 3 = −5
−3 − (+5) = −3 − 5 = −8
6 − (−4) = 6 + 4 = 10
5 + (−2) = 5 − 2 = 3

Multiplying and dividing directed numbers

Multiply and divide directed numbers as normal and then find the sign for the answer using the following rules:

- Two like signs (both + or both −) give a positive answer
- Two unlike signs (one + and the other −) give a negative answer.

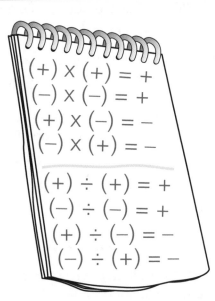

(+) × (+) = +
(−) × (−) = +
(+) × (−) = −
(−) × (+) = −

(+) ÷ (+) = +
(−) ÷ (−) = +
(+) ÷ (−) = −
(−) ÷ (+) = −

Practise the four rules of positive and negative numbers. Make one set of cards with the numbers from −10 to +10. Mix them up and turn them face down. Make a second set of cards in a different colour: three cards with the + sign, three with the − sign, three with the × sign and three with the ÷ sign. Mix them up and turn them face down on a separate part of the table. Now turn over one card from the first set and one card from the second set, followed by another from the first set. Work out the answer. So you may turn over (+9) + (−8) = 1

For example:

−6 × 3   = −18
−4 × (−2) = 8
−20 ÷ (−2) = 10
9 ÷ (−3) = −3

Factors and multiples

If you can divide one number exactly by another number, the second number is a **factor** of the first.

For example, the factors of 12 are 1, 2, 3, 4, 6, and 12.

If you multiply one number by another, the result is a **multiple** of the first number. Multiples are simply the numbers in the multiplication tables.

For example, multiples of 5 are 5, 10, 15, 20, 25, ...

Since:

$1 \times 5 = 5$
$2 \times 5 = 10$
$3 \times 5 = 15$
$4 \times 5 = 20$
$5 \times 5 = 25$

Prime numbers and factors

A **prime number** has only two factors, 1 and itself. The prime numbers up to 20 are: 2, 3, 5, 7, 11, 13, 17, 19. Note that 1 is not a prime number. Any positive integer can be written as a product of prime factors.

Example

Write 50 as a product of its prime factors.

The diagram can help you to find prime factors:

* Divide 50 by the first prime factor 2.
* Divide 25 by the prime factor 5.
  $50 = 2 \times 5 \times 5$
  $50 = 2 \times 5^2$
* Keep on going until the final number is prime.

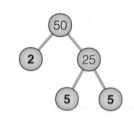

Finding the prime factors of numbers can be useful when finding the **highest common factor** (HCF) and the **lowest common multiple** (LCM) of two or more numbers.

Highest common factor

The largest factor that two numbers have in common is called the highest common factor (HCF).

Example

Find the HCF of 84 and 360.

First write the numbers as the products of their prime factors.

$84 = 2 \times 2 \times 3 \times 7$
$360 = 2 \times 2 \times 2 \times 3 \times 3 \times 5$

Ringing the factors in common gives $2 \times 2 \times 3 = 12$

HCF = 12

Lowest common multiple

The lowest common multiple (LCM) of two numbers is the lowest number that is a multiple of both numbers.

> ## Example
>
> Find the LCM of 6 and 8.
>
> $$6 = \qquad\qquad\qquad (2) \times 3$$
> $$8 = \;2\; \times \;2\; \times (2)$$
>
> 6 and 8 have a common prime factor of 2, which is only counted once.
>
> The LCM of 6 and 8 is $2 \times 2 \times 2 \times 3 = 24$

You can find the HCF and LCM of two or more values by using a Venn diagram. See Topic 8.2.

Tests of divisibility and reciprocals

To find prime numbers, simple tests of divisibility can be used.

A number is divisible by:

2 if the last digit is 0, 2, 4, 6 or 8.
e.g. 12, 48, 54
3 if the sum of the digits is divisible by 3.
e.g. 321 (3 + 2 + 1 = 6 and 6 is divisible by 3)
4 if the last two digits are divisible by 4.
e.g. 612 (12 is divisible by 4)
5 if the last digit is 0 or 5.
e.g. 25, 330
9 if the sum of the digits is divisible by 9.
e.g. 918 (9 + 1 + 8 = 18. 18 ÷ 9 = 2)

The **reciprocal** of a number $\frac{a}{x}$ is $\frac{x}{a}$.

> ## Examples
>
> The reciprocal of $\frac{4}{7}$ is $\frac{7}{4}$
>
> The reciprocal of 4 is $\frac{1}{4}$ since 4 can be written as $\frac{4}{1}$
>
> The reciprocal of $\frac{x}{2}$ is $\frac{2}{x}$
>
> To find the reciprocal of $1\frac{1}{2}$:
> - First write as an improper fraction $= \frac{3}{2}$ and then take the reciprocal.
> - Hence the reciprocal of $1\frac{1}{2}$ is $\frac{2}{3}$

Make three sets of cards numbered 0 to 9. Shuffle them. Make a two- or three-digit number. Say what the number is divisible by. Make further numbers and repeat the exercise.

Square, cube and triangular numbers

Square numbers

Square numbers are whole numbers raised to the power 2. For example:
$5^2 = 5 \times 5 = 25$ (five squared)

The first 12 square numbers are:

1	**4**	**9**	**16**	**25**	**36**	**49**	**64**	**81**	**100**	**121**	**144**
(1 × 1)	(2 × 2)	(3 × 3)	(4 × 4)	(5 × 5)	(6 × 6)	(7 × 7)	(8 × 8)	(9 × 9)	(10 × 10)	(11 × 11)	(12 × 12)

Square numbers can be illustrated by drawing squares:

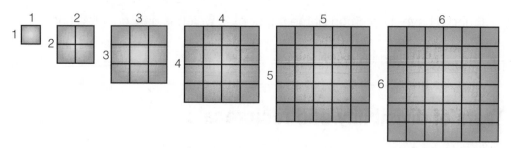

Cube numbers

Cube numbers are whole numbers raised to the power 3. For example:
$5^3 = 5 \times 5 \times 5 = 125$ (five cubed)

Cube numbers include:

1	**8**	**27**	**64**	**125**	**216**	...	**1000**
(1 × 1 × 1)	(2 × 2 × 2)	(3 × 3 × 3)	(4 × 4 × 4)	(5 × 5 × 5)	(6 × 6 × 6)	...	(10 × 10 × 10)

Cube numbers can be illustrated by drawing cubes:

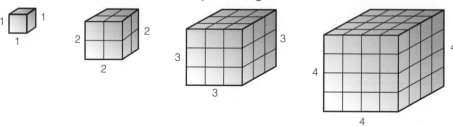

Triangular numbers

The sequence of triangular numbers is 1, 3, 6, 10, 15,...

Each time the difference between the number increases by one.

Triangular numbers can be illustrated by drawing triangle patterns:

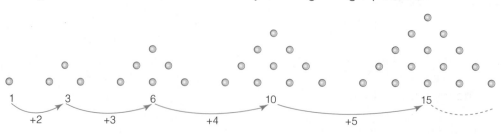

Square roots and cube roots

$\sqrt{}$ is the square root sign. Taking the square root is the opposite of squaring. When a number is square rooted it has two square roots, one positive and one negative.

For example:

$\sqrt{25} = 5$ or -5 since $(5)^2 = 25$ and $(-5)^2 = 25$

$\sqrt{196} = \sqrt{(4 \times 49)} = 2 \times 7 = 14$

| $\sqrt[3]{}$ is the cube root sign. Taking the cube root is the opposite of cubing. Examples: $\sqrt[3]{27} = 3$ since $3 \times 3 \times 3 = 27$ $\sqrt[3]{-125} = -5$ since $-5 \times -5 \times -5 = -125$ | $\sqrt[4]{}$ is known as the fourth root. Example: $\sqrt[4]{16} = 2$ since $2 \times 2 \times 2 \times 2 = 16$ | $\sqrt[5]{}$ is known as the fifth root. Example: $\sqrt[5]{243} = 3$ since $3 \times 3 \times 3 \times 3 \times 3 = 243$ |

It is important to note that $\sqrt{a} + \sqrt{b}$ is not equal to $\sqrt{a+b}$

For example, $\sqrt{9} + \sqrt{4}$ is not equal to $\sqrt{13}$

A **surd** is the square root of any number that is not a square number. It cannot be written exactly as a decimal.

For example, $\sqrt{2}, \sqrt{3}, \sqrt{5}, \sqrt{6}, \sqrt{7}, \ldots$ are all surds.

Example

Write $\sqrt{18}$ in terms of the simplest possible surd.

$$\sqrt{18} = \sqrt{9} \times \sqrt{2}$$
$$= 3 \times \sqrt{2}$$
$$= 3\sqrt{2}$$

Use a snakes and ladders board. Pick any number at random or ask a friend to choose one for you. Write down everything you know about that number, such as its multiples, factors, square root or cube root. Use the board to locate and learn all the prime, square, cube and triangular numbers from 1–100.

Indices

An **index** (plural: indices) is sometimes called a power. It can be written as:

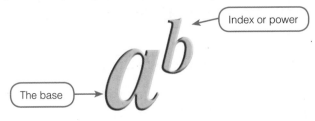

Index or power

$$a^b$$

The base

The base is the value that has to be multiplied. The index indicates how many times.

For example:

6^4 is read as '6 to the power of 4'. It means $6 \times 6 \times 6 \times 6$.
2^7 is read as '2 to the power of 7'. It means $2 \times 2 \times 2 \times 2 \times 2 \times 2 \times 2$.

Index laws

There are several laws of indices. These laws only apply when the base is the same:

Law	Examples	
When multiplying, add the powers. This also applies when the powers are negative.	$4^3 \times 4^2$	$\begin{aligned} &= (4 \times 4 \times 4) \times (4 \times 4) \\ &= 4 \times 4 \times 4 \times 4 \times 4 \\ &= 4^5, \text{ i.e. } 4^{(3+2)} \end{aligned}$
	$6^{-2} \times 6^{12}$	$= 6^{10}$
When dividing, subtract the powers. This also applies when the powers are negative.	$6^5 \div 6^2$	$\begin{aligned} &= (6 \times 6 \times 6 \times 6 \times 6) \\ &\quad \div (6 \times 6) \\ &= 6 \times 6 \times 6 \\ &= 6^3, \text{ i.e. } 6^{(5-2)} \end{aligned}$
	$8^{-4} \div 8^3$	$= 8^{-7}$
Any number raised to the power zero is just 1, provided the number is not zero. N.B. 0^0 is undefined (has no meaning).	$7^5 \div 7^5$	$= 7^{5-5} = 7^0 = 1$
	5^0	$= 1$
	2.7189^0	$= 1$
Any number raised to the power 1 is just itself.	15^1	$= 15$
	1923^1	$= 1923$
Any number raised to a negative power just turns it upside down and makes the power positive. When the index is negative always remember to take the reciprocal first.	2^{-4}	$= \dfrac{1}{2^4} = \dfrac{1}{16}$
	3^{-2}	$= \dfrac{1}{3^2} = \dfrac{1}{9}$
A fractional power is a root.	$4^{\frac{1}{2}}$	$= \sqrt{4} = 2$
	$27^{\frac{1}{3}}$	$= \sqrt[3]{27} = 3$

Standard index form

Standard index form (standard form) is a special form of index notation. It is used to write very large numbers (such as the number of stars in the Milky Way) or very small numbers (such as the diameter of an atom) in a simpler way.

When written in standard form, the number will be written as:

$$a \times 10^n$$

a must lie between 1 and 10, that is $1 \leqslant a < 10$. n is the power of 10 by which you multiply (if n is positive), or divide (if n is negative). If the number is large, n is positive; if the number is small, n is negative.

Large numbers

If the number is large, n is positive.

$634000 = 6.34 \times 10^5$

$2370 = 2.37 \times 10^3$

Small numbers

If the number is small, n is negative.

For example:

$0.00046 = 4.6 \times 10^{-4}$

$0.0361 = 3.61 \times 10^{-2}$

Calculations

The calculator can be used to do complex calculations when the numbers are in standard form.

The , , or key puts the ×10 part into the calculation.

For example:

$(2.6 \times 10^3) \times (8.9 \times 10^{12}) = 2.314 \times 10^{16}$

This is keyed in as:

And the display will usually show 2.314×10^{16} or 2.314^{16}

If carrying out a calculation involving standard form on a calculator, remember to put the ×10 part into your answer. A display of 2.314^{16} must be written as 2.314×10^{16}

When working with a calculator, the laws of indices can be used to multiply and divide numbers written in standard form.

> Make four sets of cards numbered 1 to 9. Make four zeros. Shuffle the numbered cards. Put out three or four digits followed by three or four zeros. Write down what this number will be when written in standard form. Check your answers with a calculator.

Examples

1. Work out $(2.4 \times 10^{-4}) \times (3 \times 10^7)$

$= (2.4 \times 3) \times (10^{-4} \times 10^7)$

$= 7.2 \times 10^{-4 + 7}$

$= 7.2 \times 10^3$

2. Work out $(12.4 \times 10^{-4}) \div (4 \times 10^7)$

$= (12.4 \div 4) \times (10^{-4} \div 10^7)$

$= 3.1 \times 10^{-4 - 7}$

$= 3.1 \times 10^{-11}$

1. Work out:
 a) -6×3 **b)** $12 \div -2$ **c)** $-3 - (-2)$ **d)** $6 - (-4)$

2. Say whether each statement is true or false.
 a) The factors of 20 are 1, 2, 4, 5, 10, 20.   **b)** The only even prime number is 2.
 c) The fourth multiple of 6 is 18.   **d)** The HCF of 20 and 30 is 2.
 e) The LCM of 20 and 30 is 60.   **f)** 125 is divisible by 3.
 g) The reciprocal of 7 is $\frac{1}{7}$

3. Find the value of:
 a) $\sqrt{25}$ **b)** $\sqrt[3]{64}$ **c)** $\sqrt{144}$ **d)** $\sqrt[3]{-64}$ **e)** 3^2 **f)** 4^3 **g)** 10^4

4. Write these numbers in standard form:
 a) 42 million   **b)** 632 000   **c)** 0.03210   **d)** 50047   **e)** 0.000064

5. Evaluate the following:
 a) $(2 \times 10^6) \times (3 \times 10^4)$ **b)** $(1.2 \times 10^{-9}) \div (2 \times 10^{-4})$

6. Write the following numbers in binary.
 a) 12   **b)** 23   **c)** 16

Progress Check

1.2 # Fractions and decimals

Fractions

A fraction is part of a whole number. $\frac{2}{5}$ means two parts out of five.

The top number is called the **numerator**; the bottom number is the **denominator**.

A fraction like $\frac{2}{5}$ is called a **proper fraction** because the denominator is greater than the numerator.

A fraction like $\frac{15}{7}$ is called an **improper fraction** because the numerator is greater than the denominator.

A fraction like $1\frac{5}{7}$ is called a **mixed number**.

Equivalent fractions

Equivalent fractions are fractions that have the same value. A fraction can be changed into its equivalent by either multiplying or dividing the numerator and denominator by the same number. If you divide the numerator and denominator by the same number, this is known as **cancelling**.

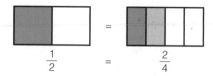

$$\frac{1}{2} \qquad = \qquad \frac{2}{4}$$

From the diagram it can be seen that $\frac{1}{2} = \frac{2}{4}$

For example:

> Fractions can be simplified if the numerator and denominator have a common factor, so you need to be able to find the common factors between two numbers. See Topic 1.1.

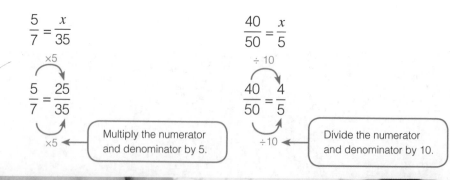

$$\frac{5}{7} = \frac{x}{35}$$

$$\frac{5}{7} = \frac{25}{35}$$

Multiply the numerator and denominator by 5.

$$\frac{40}{50} = \frac{x}{5}$$

$$\frac{40}{50} = \frac{4}{5}$$

Divide the numerator and denominator by 10.

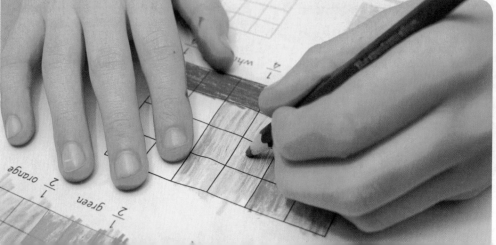

Examples

1. Write as simply as possible: $\frac{12}{18}$

$$\frac{12}{18} = \frac{2}{3}$$

since 6 is the highest common factor of 12 and 18. ← Divide both the numerator and the denominator by 6

2. Place in order $\frac{7}{8}$, $\frac{4}{5}$ and $\frac{11}{20}$, smallest first.

By converting the fractions to a common denominator, they can be easily placed in order.

The common denominator of 8, 5 and 20 is 40 (40 is the LCM of 8, 5 and 20).

$$\frac{7}{8} = \frac{35}{40}$$

$$\frac{4}{5} = \frac{32}{40}$$

$$\frac{11}{20} = \frac{22}{40}$$

In order, smallest first: $\frac{22}{40}$, $\frac{32}{40}$, $\frac{35}{40}$

Rewriting back in their original form gives: $\frac{11}{20}$, $\frac{4}{5}$, $\frac{7}{8}$

Make two sets of cards, each numbered with the digits from 1 to 20. Place one set face down and the other face up. Turn over two cards from the first set. Make a proper fraction and say whether or not it will cancel. Then see if you can make an equivalent fraction from the second set. Reverse the first fraction to make an improper fraction and convert it to a mixed number.

Adding and subtracting fractions

Only fractions with the same denominator can be added or subtracted.

Examples

1. Work out $\frac{1}{8} + \frac{3}{4}$

The lowest common denominator of 8 and 4 is 8. Replacing $\frac{3}{4}$ with $\frac{6}{8}$ gives:

$$\frac{1}{8} + \frac{6}{8} = \frac{7}{8}$$ ← Remember to only add the numerators, the denominators stay the same.

2. Work out $\frac{5}{8} - \frac{3}{16}$

$$\frac{5}{8} = \frac{10}{16}$$

$$\frac{10}{16} - \frac{3}{16} = \frac{7}{16}$$ ← 16 is the lowest common denominator of 8 and 16.

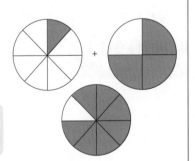

Multiplying and dividing fractions

To multiply fractions, multiply the numerators together and multiply the denominators together. Try cancelling before multiplying. Any mixed or whole numbers need to be written as improper fractions before starting.

For example:

$$\frac{2}{5} \times \frac{1}{9} = \frac{2}{45}$$

$$\frac{4}{7} \times \frac{2}{11} = \frac{8}{77}$$

$$\frac{5}{7} \times 1\frac{1}{2} = \frac{5}{7} \times \frac{3}{2} = \frac{15}{14} = 1\frac{1}{14}$$

To divide fractions, change the division into a multiplication by taking the reciprocal of the second fraction (turning it upside down) and multiplying both fractions together. If possible, cancel the fractions because it will make it easier.

Example

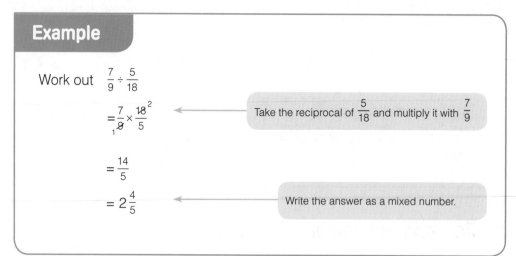

Work out $\frac{7}{9} \div \frac{5}{18}$

$$= \frac{7}{\cancel{9}_1} \times \frac{\cancel{18}^2}{5}$$

Take the reciprocal of $\frac{5}{18}$ and multiply it with $\frac{7}{9}$

$$= \frac{14}{5}$$

$$= 2\frac{4}{5}$$

Write the answer as a mixed number.

Fractions of quantities

To find a fraction of a quantity you multiply the fraction with the quantity.

Example

In a survey of 24 students, $\frac{3}{8}$ prefer English, $\frac{1}{6}$ prefer art and the rest prefer maths.

How many students prefer maths?

English $\frac{3}{8} \times 24 = 9$ $(24 \div 8 \times 3)$

Art $\frac{1}{6} \times 24 = 4$

Total $9 + 4 = 13$

Hence $24 - 13 = 11$ students prefer maths.

You also need to be able to work in reverse.

> ### Example
>
> If $\frac{1}{4}$ of a length of wood is 32cm, what is the length of wood?
>
> Let L stand for the length of wood.
>
> $\frac{1}{4} \times L = 32$
>
> $32 \times 4 = 128$cm

> You can use algebra to set up an equation for this type of problem. See Topic 3.2.

Decimals

A decimal point is used to separate whole number columns from fraction columns.

For example:

Thousands	Hundreds	Tens	Units		Tenths	Hundredths	Thousandths
6	7	1	4	•	2	3	8

Decimal point

Decimals and fractions

To change a fraction into a decimal, divide the numerator by the denominator, either by short division or by using a calculator.

To change a decimal into a fraction, write the decimal as a fraction with a denominator of 10, 100, etc. Look at the last decimal place to decide.

For example:

$\frac{2}{5} = 2 \div 5 = 0.4$

$\frac{1}{8} = 1 \div 8 = 0.125$

$0.23 = \frac{23}{100}$ ← The last decimal place is 'hundredths' so the denominator is 100.

$0.165 = \frac{165}{1000} = \frac{33}{200}$ ← The last decimal place is 'thousandths' so the denominator is 1000.

Decimals that never stop and have a repeating pattern are called **recurring decimals**. All fractions give either **terminating** or recurring decimals.

For example:

$\frac{1}{3} = 0.333\,333\,3\ldots$	usually written as $0.\dot{3}$
$\frac{5}{11} = 0.454\,545\,45\ldots$	usually written as $0.\dot{4}\dot{5}$
$\frac{4}{7} = 0.571\,428\,571\ldots$	usually written as $0.\dot{5}7142\dot{8}$

Recurring decimals can be changed into fractions.

Example

Change $0.\dot{2}$ into a fraction in its **lowest terms**.

Let $x = 0.222\,222\,222\,2\ldots$ 1
Then $10x = 2.222\,222\,222\,2\ldots$ 2

Multiply by 10^n where n is the length of the recurring pattern. In this example $n = 1$.

Subtract equation 1 from equation 2. This has the effect of making the recurring pattern disappear:
$9x = 2$
$x = \dfrac{2}{9}$ ←   Divide both sides by 9.

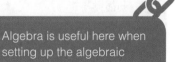

Algebra is useful here when setting up the algebraic equation. See Topic 3.2.

Ordering decimals

When ordering decimals:

- first write them with the same number of figures after the decimal point
- then compare whole numbers, digits in the tenths place, and digits in the hundredths place and so on.

Make four sets of cards numbered 0–9 and five cards with a decimal point on. Make five random numbers each with a units digit followed by a decimal point and then followed by one, two or three decimal places. Arrange these numbers in order of size from smallest to largest or from largest to smallest. You could have some extra zeros to help you as per the example.

Example

Arrange these numbers in order of size, smallest first:
5.29, 5.041, 5.7, 2.93, 5.71

First rewrite them to the same number of decimal places:
5.290, 5.041, 5.700, 2.930, 5.710

Then reorder them:
2.930, 5.041, 5.290, 5.700, 5.710

Rewrite in original form:
2.93, 5.041, 5.29, 5.7, 5.71

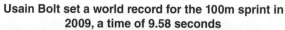

Usain Bolt set a world record for the 100m sprint in 2009, a time of 9.58 seconds

Decimal scales

Decimals are usually used when reading scales. Measuring jugs, rulers and weighing scales are examples of scales that have decimals.

For example:

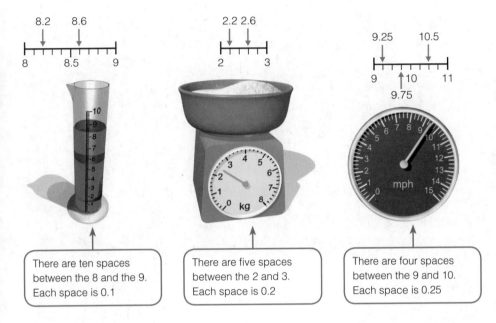

> There are ten spaces between the 8 and the 9. Each space is 0.1

> There are five spaces between the 2 and 3. Each space is 0.2

> There are four spaces between the 9 and 10. Each space is 0.25

> With an adult's permission, gather all the measuring jugs, rulers and weighing scales in your house. Practise interpreting the various scales shown on them.

Multiplying and dividing decimals by powers of 10

Remember the following rules:

Rule	Examples
To multiply decimals by 10,100 or 1000, move each digit one, two or three places to the left.	16420 × 10000 16.42 × 10 **1.642** × 100 × 1000 1642 164.2
To divide decimals by 10, 100, 1000, etc. move each digit one, two or three places to the right.	0.01735 ÷ 10000 1.735 ÷ 100 **173.5** ÷ 10 0.1735 ÷ 1000 17.35
When multiplying by a number between 0 and 1, the answer is smaller than the starting value.	0.04 × 0.01 0.4 × 0.1 **4** × 0.001 0.004
When dividing by a number between 0 and 1, the answer is bigger than the starting value.	400 ÷ 0.01 40 ÷ 0.1 **4** ÷ 0.001 4000

> Multiplying by numbers between 0 and 1 is used in probability. See Chapter 8.

> On a large piece of paper draw columns from millions to thousandths (i.e. 11 columns including one for the decimal point). Make four sets of cards from 0 to 9. You will also need some extra zeros. Make sure your columns are wide enough to fit your cards. Place two, three, four or five-digit numbers into the columns. Practise multiplying and dividing by powers of ten by sliding the digits right or left through the columns, inserting or removing zeros as necessary.

Progress Check

1. Place these fractions in order, smallest first.

$\frac{1}{2}, \frac{2}{7}, \frac{5}{14}, \frac{3}{28}$

2. Change these fractions into decimals.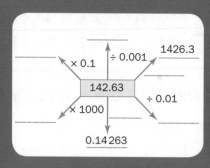

a) $\frac{5}{9}$ b) $\frac{4}{5}$ c) $\frac{6}{13}$

3. Which is the correct answer to $\frac{2}{9} \div \frac{1}{3}$?

A $\frac{2}{27}$ B $\frac{2}{3}$ C $\frac{3}{12}$ D $\frac{6}{27}$

4. In a class of 32 students, $\frac{3}{4}$ are right handed. How many students are left handed?

5. Arrange these decimals in order of size, smallest first:
0.046, 0.032, 0.471, 0.4702, 0.4694

6. Change $0.\dot{7}$ into a fraction.

7. Fill in the spaces below:

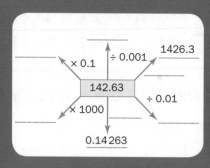

1.3 Percentages

A **percentage** is a fraction with a denominator of 100. % is the percentage sign.

For example:

75% means $\frac{75}{100}$ which is equivalent to $\frac{3}{4}$

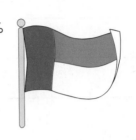

Example

A flag has three colours – red, white and blue. If 30% is red and 45% is blue, what percentage is white?

30 + 45 = 75%

Rest is white: 100 − 75 = 25%

Fractions, decimals and percentages

Fractions, decimals and percentages are different ways of expressing parts of a whole quantity.

To change a percentage into a decimal…

- first write as a fraction with a denominator of 100
- then divide the numerator by the denominator.

$13\% = \frac{13}{100} = 0.13$

To change a fraction or decimal into a percentage…

- multiply by 100.

$\frac{2}{5} \times 100 = \frac{200}{5} = 40\%$

Example

Josh sat two maths tests. He scored $\frac{16}{27}$ in Test 1 and $\frac{14}{24}$ in Test 2. Work out the percentage he got in each test, to the nearest whole number. In which test did he do the best?

Test 1

$\frac{16}{27} \times 100 = 59\%$

Test 2

$\frac{14}{24} \times 100 = 58\%$

Josh did better in Test 1.

You need to know these common fractions and their equivalents:

Fraction	Decimal	Percentage
$\frac{1}{2}$	0.5	50%
$\frac{1}{3}$	$0.\dot{3}$	$33.\dot{3}$%
$\frac{2}{3}$	$0.\dot{6}$	$66.\dot{6}$%
$\frac{1}{4}$ $\xrightarrow{1 \div 4}$	0.25 $\xrightarrow{\times 100}$	25%
$\frac{3}{4}$	0.75	75%
$\frac{1}{5}$	0.2	20%
$\frac{2}{5}$	0.4	40%
$\frac{3}{5}$	0.6	60%
$\frac{4}{5}$	0.8	80%
$\frac{1}{8}$	0.125	12.5%
$\frac{3}{8}$	0.375	37.5%
$\frac{1}{10}$	0.1	10%
$\frac{1}{100}$	0.01	1%

Make one set of cards with the values shown in the table on them. Place them face down. Turn over one card and recall the other two equivalents as quickly as you can.

Ordering

When putting fractions, decimals and percentages in order of size, it is best to change them all to decimals first.

Example

Place in order of size, smallest first:

$\frac{1}{4}$, 0.241, 29%, 64%, $\frac{1}{3}$

Put into decimals:

0.25, 0.241, 0.29, 0.64, $0.\dot{3}$

Now order:

0.241, 0.25, 0.29, 0.64, $0.\dot{3}$

Now rewrite in the original form:

0.241, $\frac{1}{4}$, 29%, $\frac{1}{3}$, 64%

Probabilities can be written as fractions, decimals or percentages. See Chapter 8.

Percentages of a quantity

The word 'of' means multiply.

When calculating a percentage of a quantity using a mental method, find 10% or 1% first.

VAT (value added tax) is charged at 20% and can be calculated mentally using the same method.

Examples

1. Find 20% of £320, without using a calculator.

   If $10\% = \frac{1}{10}$ then

   10% of 320 = $\frac{320}{10}$ = £32

   20% = 2 × 10%
   20% = 2 × 32
         = £64

2. Find 15% of £650.

   $10\% = \frac{650}{10}$ = £65

   5% is half of 65 = £32.50
   15% = 65 + 32.50
        = £97.50

When finding a percentage of a quantity with a calculator, multiply by the percentage and divide by 100.

Example

Find 12% of £20.

$20 \times \frac{12}{100}$ = £2.40

One quantity as a percentage of another

To find one quantity as a percentage of another, divide the first quantity by the second quantity and multiply by 100%.

Example

A survey showed that 42 people out of 65 preferred salt and vinegar flavoured crisps.

What percentage preferred salt and vinegar crisps?

$\frac{42}{65} \times 100 = 64.6\%$ (1 d.p.)

Make a fraction with the two numbers and then multiply by 100%.

Write down how long you spend on different activities on a Saturday (e.g. sleeping, eating, playing football, shopping, etc.) to the nearest hour, making sure the total equals 24 hours. For each activity, change these times to fractions of a day (e.g. 8 hours' sleep will equal $\frac{8}{24}$). Then, using your calculator, change the fractions into percentages, rounding any decimals to the nearest whole number. Make a 100-square grid and colour in the appropriate number of squares to show the percentage of the day spent on each activity.

Finding a percentage increase or decrease

Percentages often appear in real-life problems. If a quantity is increased by a percentage, then that percentage of the quantity is added to the original. If a quantity decreases by a percentage, then that percentage of the quantity is subtracted from the original.

An alternative approach is to use multipliers. A **multiplier** is a single number that an amount is multiplied by in order to increase or decrease that amount.

Examples

1. Two years ago the average price of a three-bedroom house was £186 000. Since then the average price of a three-bedroom house has risen by 35%. Work out the average price now.

   100% = £186 000
   Increase = 35% of £186 000
   $\frac{35}{100} \times 186\,000 = £65\,100$

   Average price is now:
   £186 000 + £65 100
   = £251 100

   Using the multiplier method:
   100% = 1, so increasing by 35% is the same as multiplying by 100% + 35% = 135%
   $135\% = \frac{135}{100} = 1.35$

   186 000 × 1.35
   = £251 100

2. Asif has £1500 in his savings account. Simple interest is paid at 4.5% p.a. How much does Asif have in his account at the end of the year?

   Simple interest is paid each year (per annum or p.a.) and is the same amount each year.

   Increasing by 4.5% is the same as multiplying by 104.5%. The multiplier is 1.045.

   1500 × 1.045 = £1567.50

   Interest paid = £67.50

   If Asif kept his money in the account for three years, he would get:

   3 × 67.50 = £202.50

   So at the end of the three years, Asif has £1500 + £202.50 = £1702.50 in his account.

To find the result of a percentage increase, multiply by:
(1 + the percentage divided by 100).

Example

A new car was bought for £8600. After two years it had lost 30% of its value. Work out the value of the car after two years.

100% = £8600
  10% = 8600 ÷ 10 = £860
  30% = 860 × 3 = £2580

Value of the car after two years:
Original – decrease
  £8600 – £2580
      = £6020

Using the multiplier method:
100% = 1, decreasing by 30% is the
same as multiplying by
100% – 30% = 70%
  70% = 0.7
  8600 × 0.7 = £6020

To find the result of a percentage decrease, multiply by 1 – the percentage divided by 100.

Profit and loss

If you buy an item, the price you pay is the cost price. If you sell the item, the price you sell it for is the selling price. Profit (or loss) is the difference between the cost price and the selling price.

The profit or loss can be written as a percentage of the original price:

$$\text{Percentage profit} = \frac{\text{profit}}{\text{original price}} \times 100\%$$

$$\text{Percentage loss} = \frac{\text{loss}}{\text{original price}} \times 100\%$$

Examples

1. A shop bought a cooker for £350. A customer later buys the cooker for £530. Find the percentage profit.

   Profit = £530 – £350
         = £180

   $\text{Percentage profit} = \frac{\text{profit}}{\text{original price}} \times 100\%$

   $= \frac{180}{350} \times 100\%$

   = 51% profit (to the nearest whole number)

2. Jackie bought a bed for £730. She later sold it for £420.

   Calculate her percentage loss.

   Loss = £730 – £420
        = £310

   $\text{Percentage loss} = \frac{310}{730} \times 100\%$

   = 42.5% loss (1 d.p.)

Repeated percentage change

Questions on repeated percentage change ask you to look at the change in value over a period of time.

Examples

1. A car was bought for £8000. Each year it depreciated in value by 20%. What was the car worth three years later? 🖩

   Method 1
   Find 80% of the value of the car first:

   100% − 20% = 80%

   Year 1: $\frac{80}{100} \times 8000 = £6400$

   Then work out the value year by year:

   Year 2: $\frac{80}{100} \times 6400 = £5120$

   Year 3: $\frac{80}{100} \times 5120 = £4096$ after three years

   Method 2
   A quick way to work this out is to use the multiplier method. Finding 80% of the value of the car is the same as multiplying by 0.8. 0.8 is the multiplier.

   Year 1: 0.8 × 8000 = £6400   ← £8000 depreciates in value by 20%
   Year 2: 0.8 × 6400 = £5120
   Year 3: 0.8 × 5120 = £4096   ← £5120 depreciates in value by 20%

   This is the same as working out $(0.8)^3 \times 8000 = £4096$, which is much quicker.

2. Jonathan has £2500 in his savings account and compound interest is paid at 4.4% per annum (per year). How much will he have in his account after three years? 🖩

   Compound interest is an example of repeated percentage change because interest is paid on the interest earned as well as on the original amount.

   Year 1:

   $1 + \frac{4.4}{100}$ is the multiplier

   1.044 × 2500 = £2610

   Year 2:
   1.044 × 2610 = £2724.84

   Year 3:
   1.044 × 2724.84 = £2844.73 (to the nearest penny)

   Total = £2844.73 (to the nearest   ← This could have been calculated as $(1.044)^3 \times 2500$.
   penny)

Reverse percentages

Reverse percentage is when the original quantity is calculated.

Example

The price of a television is reduced by 20% in the sales. It now costs £840. What was the original price?

The sale price is 100% – 20% = 80% of the original price.

$\frac{80}{100} = 0.8$

0.8 × original price = £840

Original price $= \frac{840}{0.8}$

$= £1050$

original price $\xrightarrow{\times 0.8}$ new price

$\xleftarrow{\div 0.8}$

To find the value before a percentage increase, divide by 1 + the percentage divided by 100.

To find the value before a percentage decrease, divide by 1 – the percentage divided by 100.

1. Change these percentages to fractions and decimals:
   a) 20%
   b) 32%
   c) 85%
   d) 210%

2. A meal costs £84. VAT at 20% is added to the cost of the meal. How much does the meal cost including VAT?

3. William got 62 out of 80 in a test. What percentage is this?

4. A house was bought for £165 000. Three years later it was sold for £190 000. Work out the percentage profit.

5. 15 000 people visited a museum this year. This was an increase of 20% on last year. How many visitors were there last year?

Progress Check

1.4 Ratio and proportion

Simplifying ratios

A **ratio** is used to compare two or more related quantities. 'Compared to' is replaced with two dots :

For example, '16 boys compared to 20 girls' can be written as 16 : 20. To simplify ratios, divide both parts of the ratio by the highest common factor. Here is an example:

> This is similar to cancelling fractions. See Topic 1.2.

16 : 20 = 4 : 5 ← Divide both sides by 4

Examples

1. Simplify the ratio 21 : 28 ← Divide both sides by 7

= 3 : 4

2. Express the ratio 5 : 2 in the ratio n : 1

$5 : 2 = \frac{5}{2} : \frac{2}{2}$ ← Divide both parts by 2

= 2.5 : 1

Sharing a quantity in a given ratio

To divide in a ratio:
1. Add up the total parts
2. Work out what one part is worth
3. Work out what the other parts are worth.

Example

A business makes a profit of £32 000.
The profit is divided between the directors in the ratio 3 : 2 : 5.
How much do they each receive?

3 + 2 + 5 = 10 parts
10 parts = £32 000
1 part = $\frac{32000}{10}$
1 part = £3200

So the directors get:
3 × 3200 = £9600
2 × 3200 = £6400
5 × 3200 = £16 000

Check: the total should equal £32 000

Direct and inverse proportion

Two quantities are in direct proportion if their ratios stay the same when the quantities increase or decrease.

Look at the exchange rates in the financial pages of a newspaper. Work out how many euros and dollars you could buy for different amounts of pounds sterling. Try to work them out in your head and then check your answers with a calculator.

Examples

1. A picture of length 12cm is to be enlarged in the ratio 3 : 7. What is the length of the enlarged picture?

   $12 \div 3 = 4cm$ ← Divide 12cm by 3 to get 1 part

   $4 \times 7 = 28cm$ ← Multiply 1 part by 7 to get the length of the enlarged picture

2. This is a recipe for 8 biscuits:

   80g of butter
   100g of sugar
   2 eggs
   120g of flour

   Sophie has 190g of flour. Does Sophie have enough flour to make 12 biscuits?

   There are several ways to work out this answer:

   8 biscuits   = 120g of flour

   1 biscuit $= \frac{120}{8} = 15g$

   12 biscuits $= 12 \times 15$
                $= 180g$

   Yes, Sophie has enough flour to make 12 biscuits.

   Alternatively work out the scale factor or multiplier:

   $\frac{12}{8} = 1.5$

   So, Sophie needs 1.5 times as much flour to make 12 biscuits:
   $120 \times 1.5 = 180g$

   Yes, Sophie has enough flour to make 12 biscuits.

Look at a variety of recipes for cakes and biscuits. Alter the number of cakes or biscuits you might want to make and work out the right amount of ingredients to use. Practise using multipliers as you do so.

3. At a bank, Rashna changes £80 into US $156.80. How many US dollars would Rashna get for changing £230?

   £80   = $156.80

   £1 $= \frac{156.80}{80}$

   £1   = $1.96
   £230   = 230 × 1.96

           = $450.80

Scale factors and multipliers are used in scale drawings, maps and calculating missing lengths in similar shapes. See Topic 6.1 and 5.2.

Two quantities are in inverse proportion if one increases at the same rate as the other decreases.

Example

It takes 8 people 6 days to build a wall.
At the same rate, how long does it take 3 people?

Time for 8 people = 6 days
Time for 1 person = 6 × 8 = 48 days
Time for 3 people = $\frac{48}{3}$

3 people take one-third of the time taken by 1 person

= 16 days

Best buys

Many products are sold in different-sized boxes. For example, a supermarket may have different-sized boxes of cornflakes. It is important to be able to work out which is the better value for money.

The price labels on supermarket shelves usually give you a unitary value (e.g. if a 400g jar of coffee costs £2, it will also tell you the price per gram or per 100g). Look at these on your next visit to a supermarket – it will help you gain a better understanding of ratio and proportion problems.

Example

The same brand of coffee is sold in two different-sized jars.
Which jar represents the better value for money?
Find the cost per gram for both jars:

100g costs 186p so $\frac{186}{100}$ = 1.86p per gram

250g costs 347p so $\frac{347}{250}$ = 1.388p per gram

Since the larger jar costs less per gram, it is the better value for money.

£3.47

£1.86

100g   250g

Harder proportion

The notation α means 'is directly proportional to'. This is often abbreviated to 'is proportional to'.

$y \alpha x$ means that when x is multiplied by a number, then so is the corresponding value of y. For example:

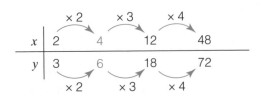

> This links to the work on straight line graphs, since $y = kx$ is in direct proportion. See Topic 4.2.

If $y \alpha x^2$, when x is multiplied by a number, y is multiplied by the square of the number.

$y \alpha \frac{1}{x}$ means that y is inversely proportional to x. When x is multiplied by a number, then y is divided by that number, and vice versa. For example:

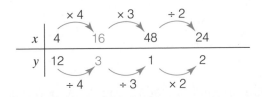

Progress Check

1. Three bars of chocolate cost £1.20. How much will four bars cost?

2. A boy spent his savings of £40 on books and DVDs in the ratio 1 : 3. How much did he spend on DVDs?

3. A map has a scale of 1 : 10 000. What distance in metres does 5cm represent on the map?

4. Complete the table:
    $y \alpha x^2$

    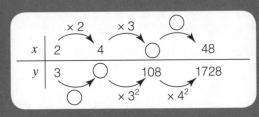

Worked questions

This is a proportional division question. The amount each person receives is proportional to the amount they paid to start with – 20 : 80 : 100 (not 20 : 80 : 1). In a ratio each part must be in the same units – in this case, pence. Now simplify the ratio. Each amount will divide by 20 giving a new ratio of 1 : 4 : 5. This equates to 10 equal parts (1 + 4 + 5 = 10).

Find the amount for one part by dividing the total amount by the number of parts.

Always check that your separate totals equal the original amount:
£60 + £240 + £300 = £600

Remember the ratio matches the order of the words, so here 3 parts of the pegs are wooden and 7 parts plastic. There are two ways to work this out.

First you can use a multiplier. 7 has been multiplied by 4 to get to 28.

Whatever you do to one side, you do to the other so multiply 3 by 4.

Alternatively, you can use the ratio as a fraction. In the basket there are wooden and plastic pegs in the ratio of 3 : 7. This means there are $\frac{3}{7}$ as many wooden as plastic.

1. James, Areeba and Diane bought a £2 lottery ticket between them. James paid 20p, Areeba paid 80p and Diane paid £1.
   They won a prize of £600 and shared it in the same ratio.
   How much did they each receive?   *(2 marks)*

   £600 ÷ 10 = £60 (= 1 part)

   James receives 1 part = £60
   Areeba receives 4 parts = 4 × £60 = £240
   Diane receives 5 parts = 5 × £60 = £300

2. The ratio of wooden pegs to plastic pegs in my peg basket is 3 : 7.
   If I have 28 plastic pegs, how many pegs are there altogether?   *(1 mark)*

   7 × 4 = 28 plastic pegs

   3 × 4 = 12 wooden pegs
   There are 12 + 28 = 40 pegs altogether.
   There are $\frac{3}{7}$ of 28 = 12 wooden pegs.
   There are 12 + 28 = 40 pegs altogether.

3. a) Write the factors of 36. *(1 mark)*

1	36
2	18
3	12
4	9
6	

The factors of 36 are 1, 2, 3, 4, 6, 9, 12, 18 and 36.

> It is easier to factorise small numbers in this way. You would use a factor tree for large numbers or when asked to write the number as a product of its prime factors.

b) Write 270 and 105 as a product of their prime factors and find the highest common factor (HCF) and the lowest common multiple (LCM). *(4 marks)*

> Make a factor tree.

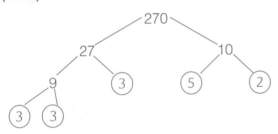

$270 = 2 \times 3 \times 3 \times 3 \times 5$

$270 = 2 \times 3^3 \times 5$

> It is usually better to write your answer in index form.

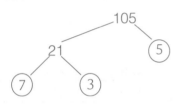

$105 = 3 \times 5 \times 7$

$HCF = 3 \times 5 = 15$

$LCM = 2 \times 3^3 \times 5 \times 7 = 1890$

$HCF = 15; LCM = 1890$

> For HCF, pick out what is common to both numbers.

> For LCM include all factors discounting those which are repeated.

4. a) The water butt in Mr Roebuck's garden holds 40 litres of water when full. He uses five-eighths to water his plants. How many litres are left? *(2 marks)*

$1 - \frac{5}{8} = \frac{3}{8}$ *is left.*

$\frac{1}{8}$ *of 40 = 40 ÷ 8 = 5*

$\frac{3}{8} = 3 \times 5 = 15$

There are 15 litres of water left in the water butt.

> Fractions questions often expect you to find the complementary amount. The full water butt is the whole one.

> Before you can find $\frac{3}{8}$ find $\frac{1}{8}$.

b) The petrol tank on Mr Roebuck's lawn mower is $\frac{3}{7}$ full when it holds 12 litres.
How many litres does it hold when full? *(2 marks)*

To find $\frac{1}{7}$, divide 12 by 3 = 4

Therefore $\frac{7}{7}$ (the whole one) is 4 × 7 = 28 litres

Mr Roebuck's petrol tank holds 28 litres.

> In this question, the amount given (12 litres) equals $\frac{3}{7}$ of the whole.

You cannot add fractions with different denominators, so you have to make equivalent fractions to give them the same denominators. Finding a common denominator is like finding the LCM. You need to find the LCM because it is always better to work with low numbers.

The lowest common denominator here is 24.

The total number of plants is the whole one: 1

c) In Mr Roebuck's herb garden, $\frac{3}{8}$ of the plants are mint, $\frac{1}{6}$ are thyme, a quarter are basil and the rest are tarragon.

What fraction of the herbs is tarragon? *(2 marks)*

$$\frac{3}{8} + \frac{1}{6} + \frac{1}{4}$$

$$\frac{9}{24} + \frac{4}{24} + \frac{6}{24}$$
$$= \frac{19}{24}$$
$$1 = \frac{24}{24}$$
$$\frac{24}{24} - \frac{19}{24} = \frac{5}{24}$$

$\frac{5}{24}$ of the herbs are tarragon.

Write the number in its numerical form – 130 000. Count the number of places the last digit on the left has to move to reach the units column. In this example it is 5.

Put each number into standard form. Now they are easy to organise and you do not need to expand them. Even though the 1 in the figure for France is the smallest unit, it will move six places as opposed to five in the other numbers. Being the same power, the other numbers are ordered in size, with the 9 in the Canada figure being the largest and the 5 in the USA figure being the smallest.

5. In an earthquake disaster, 130 thousand people lost their lives.

   a) Give this number in standard form. *(1 mark)*

   130 000

   Written in standard form, the number is 1.3×10^5

   b) Here are the amounts donated by various countries to the relief fund: *(1 mark)*

Country	Amount in £	Country	Amount in £
UK	7.6×10^5	USA	54.1×10^4
Canada	0.9×10^6	France	1.3×10^6

   Which country donated the most and which country donated the least? *(1 mark)*

   UK – £7.6×10^5 USA – £5.41×10^5

   Canada – £9×10^5 France – £1.3×10^6

   France donated the most and USA donated the least.

1% = £5000 ÷ 100. Each digit moves two places to the right = £50.

6. A building society pays 2% simple interest each year. Joe invests £5000.

   How much will he have after three years? *(2 marks)*

   Interest each year: $\frac{2}{100} \times 5000$

   2% = £100

   Interest for 3 years: £100 × 3 = £300

   £5000 + £300 = £5300

   After 3 years Joe will have £5300

Practice questions

1. These are the meter readings on Mr Patel's latest electricity bill:

ELECTRICITY BILL

CURRENT READING: 41 522.65kWh

PREVIOUS READING: 40 798.23kWh

* KWh = kilowatt hour

---

   a) What is the value of the 1 in words and figures in the present reading?   *(1 mark)*

   b) What is the value of the 3 in words and figures in the previous reading?   *(1 mark)*

   c) Write each of your answers in **a)** and **b)** in standard form.   *(2 marks)*

   d) Mr. Patel used 724.42kWh. He wants to compare his usage with that of his neighbours.

   Put these numbers in order starting with the highest:

742.04	724.8	741	728.72	742.7	724.42

   (2 marks)

2. Use = or ≠:

   a) 5.06×100 ☐ 56   *(1 mark)*

   b) $36 \div 1000$ ☐ 0.036   *(1 mark)*

   c) $1.87 \div 0.1$ ☐ 0.187   *(1 mark)*

3. Use < or > to compare these calculations.

   a) $(5 + 4 - 13)$ ☐ $(-5 - 3)$ *(1 mark)*

   b) $(-14 - -9)$ ☐ $(-14 - 9)$ *(1 mark)*

   c) $(62 - 52)$ ☐ $\sqrt{16}$ *(1 mark)*

4. a) Which of these are prime numbers?

21	31	41	51	61	71	81	91	101

   (1 mark)

   b) Using 1, 7, 13 and one of the above numbers, make two equivalent fractions.   *(2 marks)*

   c) Is 231 a prime number? Explain your answer.   *(1 mark)*

5. This table shows the heights of three objects above sea level and the depths of three objects below sea level.

Object	Height above (+) or below (−) sea level
Lighthouse	+12m
Deck of a boat	+8m
Rock	+5m
Sea level	0
Lobster pot	−3m
Fishing net	−6m
Treasure chest	−13m

a) What is the distance between the top of the lighthouse and the lobster pot? *(1 mark)*

b) What is the distance between the treasure chest and the lobster pot? *(1 mark)*

c) Which two objects have a distance of 14m between them? *(1 mark)*

d) A diver goes from the deck of the boat vertically to the treasure chest and back to the surface of the water.

   Through how many metres has he travelled? *(1 mark)*

6. a) Write the missing factors of these numbers. *(3 marks)*

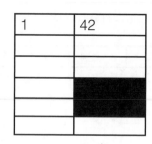

b) What is the highest common factor (HCF) of 32 and 42? *(1 mark)*

c) i) Draw a factor tree for 24 and 42 to show each number as a product of its prime factors. Write your answers in index form. *(2 marks)*

   ii) Use your answers to calculate the lowest common multiple (LCM) of 24 and 42. *(1 mark)*

7. Answer these.

a) Express $\frac{4}{5}$ as a decimal. *(1 mark)*

b) Express $\frac{4}{15}$ as a decimal. *(1 mark)*

c) Express 0.7̇6̇ as a fraction. *(1 mark)*

8. Give the reciprocal of each number.

Your answer should be written as a fraction or mixed number.

a) 6 *(1 mark)*

b) $\frac{4}{5}$ *(1 mark)*

9. Give the reciprocal of each number.

Your answer should be written as a decimal. Remember to show the correct notation for recurring decimals.

a) 8 *(1 mark)*

b) 7.5 *(1 mark)*

c) 11 *(1 mark)*

10. In a woodland: $\frac{2}{5}$ of the trees are oak; $\frac{1}{4}$ are sycamore; one-sixth are ash;
The rest of the trees are beech.

a) What fraction of the woodland is beech? *(2 marks)*

b) Change your answer to **a)** into a decimal. *(1 mark)*

11. Jane and Imran each had an identical box of chocolates.

a) i) Jane ate one-sixth of her box of chocolates and her mum ate two-fifths of what was left. Imran ate $\frac{3}{5}$ of his box and his mum ate $\frac{5}{6}$ of what was left.

Whose mum ate more chocolates, or did they eat the same amount? *(1 mark)*

ii) Explain how you know. *(3 marks)*

b) If each box contained 30 chocolates, show how your answer to **a) i)** is correct. *(3 marks)*

12. Answer these.

a) It takes three grooms two hours to muck out some horse stables.

Assuming that they all work at the same pace, how long would it take Katie on her own to muck out the same stables? *(1 mark)*

b) A sack containing 36 carrots is shared between two horses so that one gets $3\frac{1}{2}$ times as many as the other.

How many carrots do they each get? *(2 marks)*

13. Judie priced some wellington boots on the internet. In Spain they were 105.5 euros. In Canada they were 150.33 Canadian dollars. In Denmark they were 750.25 Danish kroner.

Using the exchange rates shown below, work out where it would be cheapest to buy the boots.  *(2 marks)*

£1 = 1.16 euros

£1 = 8.62 Danish kroner

£1 = 1.72 Canadian dollars

14. Mrs Shoesmith bought a new sofa costing £800. She paid a deposit of 30%.
Describe a mental method for calculating how much she still had to pay. *(1 mark)*

15. Mr Brown is a bus driver. 🖩

    a) Mr Brown's net income each month is £1800. Of this, he manages to save $2\frac{1}{2}$%.

        How much does he save each month?   *(1 mark)*

    b) Mr Brown's wife saves £280 from her monthly salary of £1600.

        What percentage of her salary does Mrs Brown save?   *(1 mark)*

16. In December 2011, Joan's salary was £19 600. She received an annual increase of 3% in January 2012, January 2013 and January 2014.

    a) i) Which is the correct calculator method for working out what Joan's salary was in January 2014?

19 600 × 1.3 × 1.3 × 1.3
19 600 × 1.03 × 1.03 × 1.03
19 600 × 1.03 × 3
19 600 × 1.3 × 4
19 600 + 300 + 300 + 300

                                                     (1 mark)

      ii) Explain your choice.   *(2 marks)*

    b) i) Over the same three-year period, Joan's investment of £15 000 in shares lost 2% in value annually. Which is the correct calculator method for working out what Joan's investment was worth at the end of the three-year period?

15 000 × 0.02 × 0.02 × 0.02
15 000 × 0.2 × 0.2 × 0.2
15 000 × 98 × 98 × 98
15 000 × 0.98 × 0.98 × 0.98
15 000 × 200 × 3

                                                     (1 mark)

      ii) Explain your choice.   *(2 marks)*

    c) Joan's husband, Charlie, bought a car for £12 000.

        What does the following calculation tell us about the value of Charlie's car over a two-year period?

        12 000 × 0.86 × 0.7   *(2 marks)*

    d) Joan bought her house in December last year. After a 13% price rise in house prices during the following year, her house was valued at £146 000.

        Write down the calculator calculation you would use to find out what Joan paid for her house.   *(1 mark)*

17. Wormer is given to horses in proportion to their weight. A horse is given 1.5ml for every 20 kilograms it weighs.

    a) Myst weighs 480kg. How much wormer should Myst be given?   *(2 marks)*

    b) Another horse, Jasmine, is given 48.75ml of wormer. How much does Jasmine weigh?

                                               (2 marks)

18. Jess mixes her horse's feed in the following proportions:

> 450 grams of oats
> 300 grams of barley
> 150 grams of sugar beet

    a) Simplify these proportions in the ratio of oats : barley : sugar beet.   *(1 mark)*

    b) Jess mixes larger quantities to last a few days. How many grams of barley will there be in 7.2kg of mixture?   *(2 marks)*

19. This triangle has an area of $1\frac{1}{5}$cm².

What is the height of the triangle?   *(2 marks)*

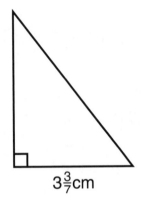

$3\frac{3}{7}$cm

2 Calculations

After studying this chapter you should be able to:
- use a variety of written and mental methods to work out calculations
- use a variety of calculator methods to work out calculations
- use a calculator efficiently and appropriately to perform complex calculations
- approximate and estimate answers to complex calculations.

 5  ## 2.1 Methods

Addition

When adding integers and decimals, the place values must line up in columns.

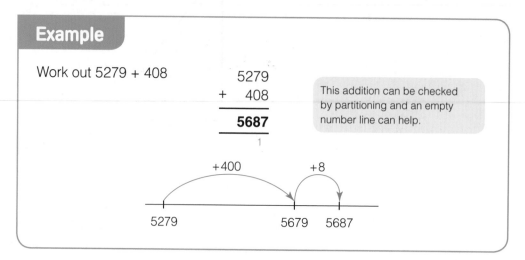

Example

Work out 5279 + 408

$$
\begin{array}{r}
5279 \\
+\quad 408 \\
\hline
\mathbf{5687} \\
{\scriptstyle 1}
\end{array}
$$

This addition can be checked by partitioning and an empty number line can help.

The same method can be used when the numbers are decimals.

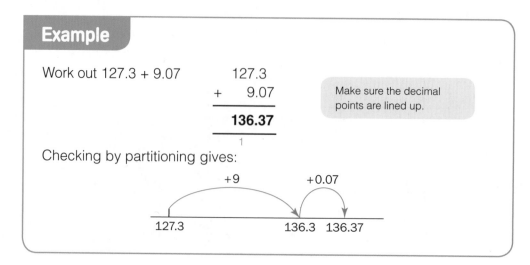

Example

Work out 127.3 + 9.07

$$
\begin{array}{r}
127.3 \\
+\quad 9.07 \\
\hline
\mathbf{136.37} \\
{\scriptstyle 1}
\end{array}
$$

Make sure the decimal points are lined up.

Checking by partitioning gives:

Subtraction

When subtracting integers and decimals, the place values must line up one on top of the other. Subtracting is also known as finding the **difference**.

> ## Example
>
> Work out 2791 – 365
>
> $$\begin{array}{r} {}^{81}27\!\!\not9\!\!\not1 \\ -\ \ \ 365 \\ \hline \mathbf{2426} \end{array}$$
>
> > In the units column 1 – 5 won't work. Borrow 10 from the next column. So the 9 becomes 8 and the 1 becomes 11.
>
> **Compensation** can be used to check the answer, by adding or subtracting too much and then compensating.
>
>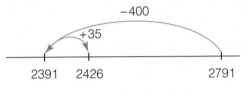
>
> −400
> +35
> 2391   2426   2791

Multiplication

Multiplication is much easier if you know your multiplication tables. Using a grid method can sometimes help when multiplying, although you do need to know how to do formal multiplication.

> ## Example
>
> Work out 6.24 × 8
>
> $$\begin{array}{r} 6.24 \\ \times\ \ \ \ 8 \\ \hline \mathbf{49.92} \\ {\scriptstyle 1\ \ 3} \end{array}$$
>
> > Multiply each of the digits 6, 2, 4 by 8. Start from the right and move to the left.
>
> Alternatively if a grid method is used:
>
> | × | 6 | 0.2 | 0.04 |
> |---|---|-----|------|
> | 8 | 48 | 1.6 | 0.32 |
>
> 48 + 1.6 + 0.32 = 49.92

Multiplying two or more numbers together is known as finding the **product**.

Make four sets of cards, each numbered with the digits 0–9. Turn them face down. Turn over two cards and say the multiplication fact and its answer. For example, you might turn over a 7 and an 8, so you would say seven eights are 56. Then practise your division facts by turning over two cards to make a two-digit number and state two numbers which multiply together to make that number. For example, you might turn over 7 and 2 to make 72. You would then say eight nines or six twelves are 72.

How would you spend £1 million? Find out the prices of items you would buy if you had £1m to spend. Then play around with the different amounts. Practise addition by choosing four or five to total. Then practise subtraction by taking these totals away from £1m. Practise multiplication by imagining that you might buy six or seven of one particular item. Find the average price of several items to practise division.

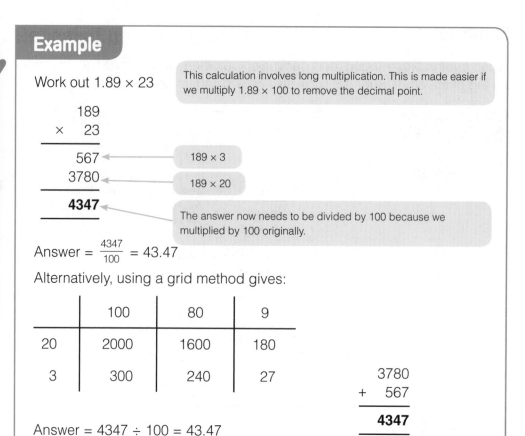

Example

Work out 1.89 × 23

> This calculation involves long multiplication. This is made easier if we multiply 1.89 × 100 to remove the decimal point.

$$
\begin{array}{r}
189 \\
\times \quad 23 \\
\hline
567 \\
3780 \\
\hline
\textbf{4347} \\
\hline
\end{array}
$$

189 × 3

189 × 20

> The answer now needs to be divided by 100 because we multiplied by 100 originally.

Answer = $\frac{4347}{100}$ = 43.47

Alternatively, using a grid method gives:

	100	80	9
20	2000	1600	180
3	300	240	27

$$
\begin{array}{r}
3780 \\
+ \quad 567 \\
\hline
\textbf{4347} \\
\hline
\end{array}
$$

Answer = 4347 ÷ 100 = 43.47

Division

Take care not to miss out important zeros in long and short division.

Ask at home if you can look at some utility bills. Try to find examples of where the four rules of number (+,−, × and ÷) have been used.

Example

A bar of chocolate costs 74p. Tracey has £9.82 to spend. What is the maximum number of bars Tracey can buy? How much change does she have left?

The method of chunking can be used when dividing. Always try to estimate the answer to your division.

Long division

$$
\begin{array}{r}
1\,3 \\
74\overline{)98\,2} \\
74- \\
\hline
242 \\
222- \\
\hline
20 \\
\end{array}
$$

98 ÷ 74 = 1 remainder 24

now bring down the 2

242 ÷ 74 = 3 remainder 20

Tracey can buy 13 bars and has 20p left over.

Chunking (here we take off multiples of 74)

$$
\begin{array}{r}
74\overline{)98\,2} \\
740- \\
\hline
242 \\
222- \\
\hline
20 \\
\end{array}
$$

74 × 10

74 × 3

Answer = 10 + 3 = 13 bars and 20p left over.

In this example, the 74 is the divisor, the 13 is the quotient and the remainder is 20p.

If the question had simply been 982 ÷ 74, then the answer could be written as $13\frac{20}{74}$ or $13\frac{10}{37}$

When dividing by decimals, it is useful to change to an equivalent calculation that does not have a decimal divisor.

For example:
- 372.8 ÷ 0.4 is equivalent to 3728 ÷ 4
- 527.1 ÷ 0.02 is equivalent to 52 710 ÷ 2

This skill is used when estimating calculations. See Topic 2.2.

Order of operations

BIDMAS is a made-up word that helps you to remember the order in which calculations take place.

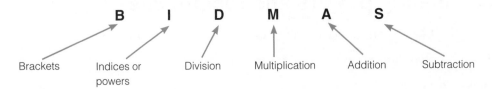

B   **I**   **D**   **M**   **A**   **S**

Brackets   Indices or powers   Division   Multiplication   Addition   Subtraction

This means that brackets are worked out first and, in the absence of brackets, division and multiplication are done before addition and subtraction.

For example:
- $(3 + 4) \times 5 = 35$ ← The brackets are carried out first.
- $5^2 - (2 \times 3) = 25 - 6 = 19$
- $6 + 3 \times 2 = 12$ ← The multiplication is carried out first.

The order of operations applies to algebra as well as all numerical calculations. See Topic 3.1.

Important calculator keys

This calculator shows some of the important calculator keys. Make sure you know how your calculator works.

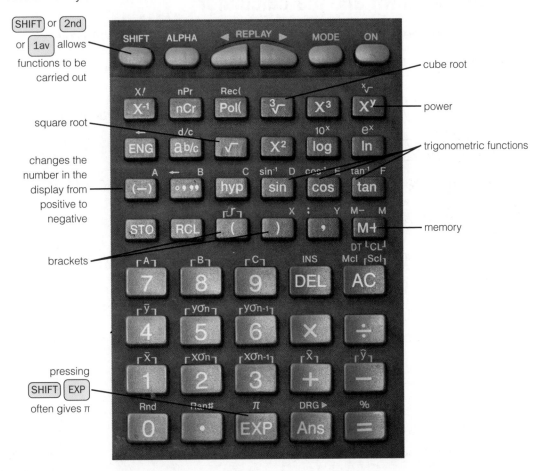

SHIFT or 2nd or 1av allows functions to be carried out

cube root

power

square root

changes the number in the display from positive to negative

trigonometric functions

memory

brackets

pressing SHIFT EXP often gives π

For example:

$$\frac{15 \times 10 + 46}{9.2 \times 2.1} = 10.14 \text{ (2 d.p.)}$$

Depending on your model of calculator, this may be keyed in as:

(15 × 10 + 46) ÷ (9 . 2 × 2 . 1) =

The above calculation can also be done using the memory keys. Try writing down the key sequence for yourself.

Calculating powers and reciprocals

y^x x^y or $x^{\bullet}$ are used to calculate powers such as 2^7.

Use the power key on the calculator to work out 2^7.

Keying in gives:

2 x^y 7 =

(Check your answer is 128)

$\frac{1}{x}$ is the reciprocal key on the calculator. It is used to calculate the reciprocal of a number.

For example:

- $9^{\frac{1}{3}} \times 4^5 = 2130$ (to the nearest whole number)
- The reciprocal of 0.3 = $3.\dot{3}$

Standard form and the calculator

To key a number in standard form into the calculator, use the ×10ˣ key.

Some calculators use EE or EXP. Check your calculator.

For example:

Depending on your model of calculator, 6.23×10^6 may be keyed in as:

6 . 2 3 ×10ˣ 6

4.93×10^{-5} can be keyed in as:

4 . 9 3 ×10ˣ − 5

Some calculators do not show standard form correctly on the display. For example:
- 7.632^7 means 7.632×10^7
- 4.2^{-09} means 4.2×10^{-9}

It is important to put in the ×10 part when you write your answer.

Practise these calculator techniques over and over again to make sure you are using your calculator correctly. Then practise some standard form calculations using your own numbers.

Interpreting the calculator display

When calculations involve money, remember:

- A display of 4.2 means £4.20 (four pounds twenty pence).
- A display of 3.07 means £3.07 (three pounds and seven pence).
- A display of 0.64 means £0.64 or 64 pence.
- A display of 6.2934 means £6.29, i.e. it has to be rounded to 2 d.p.

1. Work out the following, without using a calculator.
   a) 27.4×32 d) 3729×46
   b) $3762 \div 3$ e) $237.2 \div 0.8$
   c) $690 \div 15$

2. Work out the following on your calculator.

   a) $\dfrac{27.1 \times 6.4}{9.3 + 2.7}$ b) $\dfrac{(9.3)^4}{2.7 \times 3.6}$

   c) $\sqrt{\dfrac{25^2}{4\pi}}$ d) $\dfrac{5}{9}(25 - 10)$

3. Which is the correct answer?
   a) $2 + 3 \times 7$ **A** 35   **B** 23
   b) $4 - 1 \times 5$ **A** 15   **B** −1
   c) $(9 + 1)^2 \times 4$ **A** 1600   **B** 400

4. Erin worked out $\dfrac{5.79 + 3.27}{6.3^2 \times 4}$ on her calculator. She got 6.12 to 2 decimal places.
   a) Work out the calculation.
   b) Explain the mistake Erin made in obtaining her answer.

Progress Check

2.2 Rounding and estimating 🎧 6

Rounding numbers

Large numbers are often **approximated** to the nearest 10, 100 or 1000 to make them easier to work with.

Rounding to the nearest 10

To round to the nearest 10, look at the digit in the units column. If it is less than 5, round down. If it is 5 or more, round up.

> **Example**
>
> 568 people attended a concert. Round this to the nearest 10.
>
> There is an 8 in the units column, so round up to 570.
>
> 568 is 570 to the nearest 10.

Rounding to the nearest 100

To round to the nearest 100, look at the digit in the tens column. If it is less than 5, round down. If it is 5 or more, round up.

> **Example**
>
> In May, 2650 people went to the zoo. Round this to the nearest 100.
>
> Since there is a 5 in the tens column, we round up to 2700.
>
> 2650 is 2700 to the nearest hundred.

Rounding to the nearest 1000

To round to the nearest 1000, look at the digits in the hundreds column. The same rules apply as before.

> **Example**
>
> Round 16 420 to the nearest 1000. There is a 4 in the hundreds column, so round down to 16 000.
>
> 16 420 is 16 000 to the nearest thousand.

Use four sets of cards, each numbered with the digits 0–9, placed face down. Turn individual cards over randomly to make four or five-digit numbers. Round the numbers you create to the nearest 10, 100 and 1000.

Similar methods can be used to estimate any number to any power of 10.

Newspaper reports often use rounded numbers.

Decimal places

It is sometimes useful to round decimals to the nearest whole number or to a specified number of decimal places.

The same rules of rounding shown above are used.

To round to the nearest whole number, look at the number in the first decimal place:
- If it is 5 or more, round the units up to the next whole number.
- If it is less than 5, the units stay the same.

In many calculations involving measurements, the values are often rounded. See Topic 7.1.

> **Example**
>
> Round these to the nearest whole number:
>
> **a)** 12.3   ←—— 3 is less than 5, so round down
>
> This is 12 to the nearest whole number.
>
> **b)** 7.9   ←—— 9 is more than 5, so round up
>
> This is 8 to the nearest whole number.

To round to the nearest tenth (or to 1 decimal place), look at the number in the second decimal place:

- If it is 5 or more, round the first decimal place up to the next tenth.
- If it is less than 5, the first decimal place remains the same.

> ### Example
>
> Round these numbers to 1 decimal place.
>
> **a)** 9.45
> This is 9.5 to 1 decimal place.   ← 4 is rounded up to 5 here because the second decimal place is 5.
>
> **b)** 12.62
> This is 12.6 to 1 decimal place.   ← Stays the same because 2 is less than 5.

A similar method can be used when rounding any number to a particular number of decimal places. For example:

- $16.59 \rightarrow 16.6$ (1 d.p.)
- $12.3642 \rightarrow 12.364$ (3 d.p.)
- $8.435 \rightarrow 8.44$ (2 d.p.)

When calculating with money, we always round to 2 decimal places.

For example, £16.58 ÷ 4 = £4.145 but this would round to £4.15

Significant figures

The rule for rounding to a given number of **significant figures** is the same as for decimal places – if the next digit is 5 or more, round up.

The first significant figure is the first digit that is not a zero. The second, third and fourth significant figures follow on after the first digit. They may or may not be zero.

For example:

7.021 has 4 significant figures.   0.003 706 also has 4 significant figures.

It is important that the place value is not changed when rounding. Here are some examples of rounding to a certain number of significant figures:

Number	to 3 s.f.	to 2 s.f.	to 1 s.f.
4.207	4.21	4.2	4
4379	4380	4400	4000
0.006 209	0.006 21	0.0062	0.006

After rounding, the end zeros must be filled in.

For example, 4380 = 4400 to 2 s.f. (not 44).

No extra zeros should be put in after the decimal point.

For example, 0.013 = 0.01 to 1 s.f., not 0.010

Use four sets of cards, each numbered with the digits 0–9, placed face down. Create a further card with a decimal point on it. Turn over one of the cards, place the decimal point to the right of it and turn over one more card, placing that to the right of the decimal point. Round your number to the nearest whole number. Create numbers with 2, 3 or 4 decimal places and practise rounding these to 1, 2 or 3 decimal places.

Possible error of half a unit when rounding

If a measurement has been rounded, the actual measurement lies within a maximum of half a unit of that amount. It can be half a unit bigger or smaller.

There are two types of measurement – **discrete** and **continuous**.

Discrete measures

Discrete measures are quantities that can be counted, such as people. For example, a school has 1400 students to 2 significant figures (i.e. the nearest 100). The actual figure could be anything from 1350 to 1449.

Continuous measures

Continuous measures are measurements that have been made by using a measuring instrument, such as a person's height or weight. Continuous measures are not exact.

For example, Nigel weighs 72kg to the nearest kilogram. His actual weight (w) could be anywhere between 71.5kg and 72.5kg.

We use different calculations on discrete and continuous data when finding the average and spread, in addition to representing the data graphically. See Chapter 9.2.

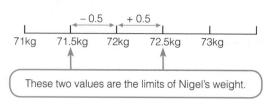

71kg   71.5kg   72kg   72.5kg   73kg

These two values are the limits of Nigel's weight.

If w represents weight, then:

$$71.5 \leqslant w < 72.5$$

This is the lower limit of Nigel's weight, sometimes known as the lower bound. Anything below 71.5 would be recorded as 71kg.

This is the upper limit, sometimes known as the upper bound of Nigel's weight. Anything from 72.5 upwards would be read as 73kg.

Measure some objects in your home, draw them and write their measurements rounded to the nearest centimetre and to the nearest 10cm. Similarly, weigh different items, draw them and write their weights rounded to the nearest gram and to the nearest 10g. Look at some of your rounded measurements and work out the minimum and maximum limits.

Example

The length of a seedling is measured as 3.7cm to the nearest tenth of a centimetre. What are the upper and lower limits of the length (l)?

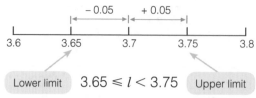

3.6   3.65   3.7   3.75   3.8

Lower limit $3.65 \leqslant l < 3.75$ Upper limit

Rounding sensibly in calculations

When solving problems the answers should be rounded sensibly. It is wise to go back and check the context of the question.

> **Examples**
>
> 1. Mr Singh organised a trip to the theatre for 420 students and 10 teachers. If a coach can seat 53 people, how many coaches did he need?
>
> $430 \div 53 = 8.11$ coaches
>   9 coaches are needed.
>
>   > Obviously, not everybody can get on 8 coaches ($8 \times 53 = 424$), so we need to round up to 9 coaches.
>
> 2. Work out 95.26×6.39
>
> $95.26 \times 6.39 = 608.7114$
> $= 608.71$ (2 d.p.)
>
>   > The answer is rounded to 2 d.p. because the values in the question are to 2 d.p.
>
> 3. James has £9.37. He divides it equally between 5 people. How much does each person receive?
>
> $£9.37 \div 5 = £1.874$
> $= £1.87$
>
>   > This is rounded to 2 d.p. because it involves money.

Checking calculations

A calculation can be checked by carrying out the inverse operation. For example:

Calculation	Inverse Operation
$106 \times 3 = 318$	$318 \div 106 = 3$ or $318 \div 3 = 106$

$$106 \quad \xrightarrow{\times 3} \quad 318$$
$$106 \quad \xleftarrow{\div 3} \quad 318$$

Calculation	Inverse Operation
$\sqrt{5} = 2.236067977$	$(2.236067977)^2 = 5$

$$5 \quad \xrightarrow{\sqrt{}} \quad 2.234067977$$
$$5 \quad \xleftarrow{x^2} \quad 2.234067977$$

A calculation can also be checked by carrying out an equivalent calculation:

$692 \times 4 = 2768$

Check with $(700 - 8) \times 4 = 2800 - 32$

> $692 = 700 - 8$, so this is an equivalent calculation.

$$= 2768$$

Estimating

Estimating or **approximating** is a good way of checking answers. Estimating can help you to decide whether an answer is the right order of magnitude, which means 'about the right size'.

When estimating:
- round the numbers to 'easy' numbers, usually one or two significant figures
- work out the estimate using these 'easy' numbers
- use the symbol ≈ which means 'approximately equal to'.

For multiplying or dividing, never approximate a number to zero. Use 0.1, 0.01, 0.001, etc.

Examples

1. Estimate the following calculations:

   a) $8.93 \times 25.09 \approx 10 \times 25 = 250$

   b) $(6.29)^2 \qquad \approx 6^2 = 36$

   c) $\dfrac{296 \times 52.1}{9.72 \times 1.14} \approx \dfrac{300 \times 50}{10 \times 1} = \dfrac{15000}{10} = 1500$

   d) $0.096 \times 79.2 \approx 0.1 \times 80 = 8$

2. Jack does the calculation $\dfrac{9.6 \times 103}{(2.9)^2}$

   a) Estimate the answer to this calculation, without using a calculator.
   $$\frac{9.6 \times 103}{(2.9)^2} \approx \frac{10 \times 100}{3^2} = \frac{1000}{9} \approx \frac{1000}{10} = 100$$

   b) Jack's answer is 1175.7. Is this the right order of magnitude?

   Jack's answer is not the right order of magnitude (size). It is 10 times too big.

There are different ways of finding an approximate answer. For example:

$8.93 \times 25.09 \approx 10 \times 25 = 250$

$\qquad\qquad$ or $9 \times 25 = 225$

In this case, 9×25 is a closer approximation.

You need to be able to recognise what makes a 'good approximation'. When adding or subtracting, very small numbers may be approximated to zero.

Example

Estimate the following calculations:

a) 109.2 + 0.0002

$\approx$ 110 + 0 = 110

b) 63.87 – 0.01

$\approx$ 64 – 0 = 64

1. Round the following numbers to the nearest 10.
   a) 268
   b) 1273
   c) 42 956
   d) 2385

2. Round the following numbers to 2 decimal places.
   a) 47.365
   b) 21.429
   c) 15.3725

3. Round the following numbers to 2 significant figures.
   a) 1247
   b) 0.003 729

Progress Check

4. Paint is sold in 8-litre tins. Sandra needs 27 litres of paint.

   How many tins must she buy?

5. Estimate the answers to the following:
   a) $\dfrac{29.4^2 + 106}{2.2 \times 5.1}$

   b) $\dfrac{521 + 207}{0.43}$

6. Mr Johnson tries to work out the answer to 292 × 42. He works it out as 1226.4, but he thinks he has made a mistake.

   a) Make a rough estimate of 292 × 42.

   b) Compare your estimate with Mr Johnson's answer. Do you think he made a mistake?

Worked questions

1. In 5 days, a pet shop sold 9.5kg, 17.37kg, 14.625kg and 19 000g of dog biscuits.

What total weight of dog biscuits was sold in five days? Give your answer in kilograms. *(1 mark)*

```
14.625        14.625
17.37         17.370
 9.5           9.500
+19          +19.000
              60.495
```

In 5 days, the pet shop sold 60.495kg of dog biscuits.

2. How much profit did a car dealer make if he bought a car for £5984.79 and sold it for £10 000? *(1 mark)*

```
10 000.00      £10 000.00
 -5984.79     -£ 5 984.79
               £ 4 015.21
```

The car dealer made £4015.21 profit.

3. a) The rim of a wheel is 3.14 times the diameter of the wheel. Find the distance around the rim if the diameter is 8.7cm. Give your answer in centimetres, rounded to 2 decimal places. *(2 marks)*

3.14 × 100 = 314 and 8.7 × 10 = 87, so this is an increase of 100 × 10 = 1000 in total

```
    314
   × 87
   2198
  25120
  27318
```

27318 ÷ 1000 = 27.318

The rim of the wheel measures 27.32cm.

b) How many complete revolutions does a wheel with a circumference of 4.3m make in covering a distance of 110.08m? *(1 mark)*

$43 = 4.3 \times 10$

$$
\begin{array}{r}
25.6 \\
43 \overline{\smash{)}1100.8} \\
86 \\
\hline
240 \\
215 \\
\hline
258 \\
258 \\
\hline
\end{array}
$$

The wheel makes 25 complete revolutions.

The calculation is 110.08 ÷ 4.3. In division of decimals, only the divisor needs to be a whole number.

To make an equivalent sum you must also multiply 110.08 by 10.

110.08 ÷ 4.3 = 1100.8 ÷ 43, so you do not need to balance at the end.

4. The crowd watching a football match last week was 5700, rounded to the nearest 100.

This is a discrete measure. The rounded unit is 100.

a) What is the smallest number of people that could have been present? *(1 mark)*
The smallest number of people that could have been present is 5650.

The minimum value is 5650 5649 or less would have been rounded to 5600.

b) What is the largest number of people that could have been present? *(1 mark)*
The largest number of people that could have been present is 5749.

The maximum value is 5749. 5750 or more would have been rounded to 5800.

5. A lawn measures 5.6m by 4.7m to the nearest 0.1m. What are the upper and lower limits of the following?

These are continuous measures. The rounded unit is one tenth.

a) The lawn's length. *(1 mark)*
Upper limit = 5.65m
Lower limit = 5.55m

b) The lawn's width. *(1 mark)*
Upper limit = 4.75m
Lower limit = 4.65m

c) The lawn's area. *(1 mark)*
Upper limit = 5.65m × 4.75m
= 26.84m²
Lower limit = 5.55m × 4.65m
= 25.81m²

d) The lawn's perimeter. *(1 mark)*
Upper limit = (5.65 + 4.75) × 2
= 20.8m
Lower limit = (5.55 + 4.65) × 2
= 20.4m

6. p and q are continuous values rounded to the nearest 10.

$p = 200$ and $q = 420$

Work out the greatest possible value of $p \div q$. Round the answer to 3 significant figures. *(2 marks)*

p = 195 − 205
q = 415 − 425
205 ÷ 415 = 0.493 975 9
= 0.494 (to 3 s.f.)

Do not assume that the largest limits will give the greatest overall value. Dividing by a smaller number gives a larger answer. If you are in doubt, either work out the answer using all four possibilities or substitute complex numbers with simple ones. For example, 12 ÷ 2 gives a larger answer than 12 ÷ 4

Practice questions

1. A builder buys a house for £75 999. He wants to make at least £15 950 profit. (▦)

   a) What is the least amount the builder could sell the house for in order to realise this profit?   *(1 mark)*

   b) Before the builder can sell the house, he has various jobs to do. Here are the quotations from the plumbers and electricians.

Plumbing	
Joe Tap	The work will take 42 hours. Charge = £12.50 per hour
Jill Washer	The work will take 42 hours. Charge = £96 per day. I work an 8-hour day.
Electrics	
Jason Watt	The work will take 38 hours. Charge = £13.75 per hour
Jamilla Spark	The work will take 38 hours. Charge = £112 per day. I work an 8-hour day.

   Calculate the amount of each quotation.

   Which plumber and which electrician should the builder choose to keep his expenses to a minimum?   *(3 marks)*

   c) The money spent on the plumber and electrician has to be taken from the builder's profit.

   Assuming he chooses the plumber and electrician you identified in **b)**, what will the builder's profit now be?   *(2 marks)*

2. Magazines are put into bundles of 38 for delivery to the newsagents.

   a) How many magazines are needed for 45 bundles?   *(1 mark)*

   b) What information will this calculation provide?
      2058 ÷ 38   *(1 mark)*

   c) Work out the calculation in **b)**.

      How many more magazines are needed to make another bundle?   *(2 marks)*

3. a) Round each number to the required number of significant figures (s.f.).
      i) 58 400 (1 s.f.)   *(1 mark)*
      ii) 0.078 51 (2 s.f.)   *(1 mark)*

   b) By rounding to 1 s.f., estimate:
      $$\frac{82 \times 2.8}{0.437 + 0.396}$$ *(2 marks)*

4. Choose the correct answer to each calculation by ticking the box next to the correct answer.

   a) $76 - (9 \times 8)$ =   536 ☐   4 ☐   *(1 mark)*

   b) $6 + 5 \times 2$ =   16 ☐   22 ☐   *(1 mark)*

   c) $8 \times (9 - 2 + 4)$ =   88 ☐   24 ☐   *(1 mark)*

   d) $(3 + 1)^2 \times 4$ =   40 ☐   64 ☐   *(1 mark)*

5. **a)** An aeroplane is carrying 230 people rounded to the nearest 10.
What is the greatest number of people that could be on the aeroplane? *(1 mark)*

b) The aeroplane is travelling at 500km per hour to the nearest 10km.
What is the aeroplane's highest possible speed? *(1 mark)*

6. A patio is measured as being 8.4m × 5.8m to the nearest 10cm.

a) What information does this calculation give?
8.45 × 5.85 *(1 mark)*

b) What information does this calculation give?
(8.35 + 5.75) × 2 *(1 mark)*

7. A rectangle measures 15m by 36m to the nearest metre.

What are the upper and lower limits of its perimeter? *(2 marks)*

8. In Jared's money box there are 150 pennies rounded to the nearest ten.

What is the largest and smallest number of coins that can be in the box? *(2 marks)*

9. x and y are continuous values rounded to the nearest unit. 🔳

$x = 25$ and $y = 19$

a) Work out the greatest value of $x - y$. *(2 marks)*

b) Work out the smallest value of $x \div y$. Round your answer to 3 significant figures. *(2 marks)*

10. **a)** Maria carried out the following calculation on her calculator: 🔳
$$\frac{(25 + 56)}{(32 \times 2)}$$

Her answer was 28.5

What did Maria do wrong? *(1 mark)*

b) Sanya used her calculator to add the costs of the sweater and trousers.

£16.82

£25.68

She was unsure how to interpret the answer. How would you explain the answer? *(1 mark)*

11. Look at this calculation:
$$\frac{(42 \times 5.86)}{(0.12 + 0.24)}$$

a) By rounding, find an approximate answer to the above calculation. *(1 mark)*

b) Find the exact answer correct to 2 decimal places without using a calculator.
Check this answer with your estimate. *(2 marks)*

Learning Summary

After studying this chapter you should be able to:
- construct and use formulae from mathematics and other subjects
- manipulate algebraic formulae, equations and expressions
- construct and solve linear equations
- solve simultaneous linear equations in two variables
- solve inequalities and find the solution set.

3.1 Symbols and formulae

Algebraic symbols

In algebra we use letters as symbols. The letters represent:
- unknown numbers in an **equation**
- **variables** in a **formula**, which can take many values, e.g. $V = IR$
- numbers in an identity, which can take any values, e.g. $3(x + 2) \equiv 3x + 6$, for any value of x.

A **term** is a number or a letter, or a combination of both multiplied together. Terms are separated by + and − signs. Each term has a + or − sign attached to the front of it.

For example:

$$5ab + 2c - 3c^2 + 5$$

Invisible + sign ab term c term c^2 term   Number term

A collection of terms is known as an **expression**.

Using letter symbols

Make a set of cards with each of the expressions from the left column of the table. Make another set, preferably on another colour, using the expressions in the second column. Turn the first set face down and the second set face up. Turn over a face-down card and see how quickly you can match it with a face-up card.

There are several rules to follow when writing expressions:

Expression	Should be written as...
$a + a + a$	$3a$
$a \times b$	ab
$a \times 3 \times b$	$3ab$
$b \times b$	b^2
$b \times b \times b$	b^3
$n \times n \times 3$	$3n^2$
$a \times (b + c)$	$a(b + c)$
$(a + b) \div c$	$\dfrac{a + b}{c}$

b^2 is not the same as $2b$

b^3 is not the same as $3b$

$3n^2$ is not the same as $(3n)^2$

Example

In a game John has r counters. Write down the number of counters each person has using r.

a) Carol has twice as many counters as John.
Carol has $2r$.

b) Vali has 12 fewer than John.
Vali has $r - 12$.

c) Stuart has half as many as John.
Stuart has $r \div 2 = \frac{r}{2}$

d) Hilary has 5 fewer than Carol.
Hilary has $2r - 5$.

Know the words

The following words are used in algebra.

Word	Meaning	Example
Expression	Any arrangement of letter symbols and numbers.	$2a + 3b - 4$
Formula	Connects two expressions containing variables, the value of one variable depending on the values of the others. It must have an equals sign.	$v = u + at$ When the values of u, a and t are known, the value of v can be found.
Equation	Connects two expressions involving definite unknown values. It must have an equals sign.	$x + 2 = 5$
Identity	Connects expressions involving unspecified numbers. An identity always remains true, no matter what numerical values replace the letter symbols. It has an $\equiv$ sign.	$3(x + 2) \equiv 3x + 6$ This is true no matter what values of x are used.
Function	A relationship between two sets of values, such that a value from the first set maps onto a unique value in the second set.	$y = 4x + 2$ For any value of x, the value of y can be calculated.

Collecting like terms

Expressions can be simplified by collecting like terms. Like terms have the same letters and powers.

> ### Example
>
> Simplify these expressions.
>
> **a)** $3a + 4a$
> $= 7a$
>
> **b)** $6a + 2b$
>
> This cannot be simplified, since there are no like terms.
>
> **c)** $3n + 2n - 4n$
> $= n$
>
> **d)** $5a + 4b + 3a - 6b$
> $= 8a - 2b$
>
> **e)** $5xy + 2yx$
> $= 7xy$ Since xy is the same as yx
>
> **f)** $5n^2 + 2n + 3n^2$
> $= 8n^2 + 2n$

Add the a terms then the b terms. The minus is part of the $6b$.

Remember to put the sign between the terms, i.e. $8a - 2b$ not $8a\ 2b$.

Use two sets cards, each numbered with the digits 2–9. Make another set including three as, three bs and three cs. Place the first set face down on the left of the table and the second face down on your right. Turn over one of each to make an expression, e.g. $4b$. Make a few more random expressions and then practise gathering like terms. You could then place + or − between them and practise adding and subtracting.

Multiplying letters, numbers and brackets

Algebraic expressions that are multiplied together can often be simplified, e.g. $5a \times 2b = 10ab$.

> ### Example
>
> Simplify these expressions.
>
> **a)** $3a \times 4b$
> $3 \times 4 \times a \times b = 12ab$
>
> **b)** $5a \times 3b \times 2c$
> $5 \times 3 \times 2 \times a \times b \times c = 30abc$
>
> **c)** $2a \times 3a$
> $2 \times 3 \times a \times a = 6a^2$

Remember $a \times a = a^2$

Repeat the last activity but replace + and − with ×. Start off by making just two expressions and then try multiplying three. You will need to add a square and cube sign just in case you turn over 2 or 3 of the same letter.

When multiplying out single and double brackets, use the same rules as for operations with numbers.

Multiplying out single brackets

Everything inside the bracket must be multiplied by everything outside the bracket. Partitioning can be used. Expand just means multiply out the brackets.

We used partitioning when carrying out number calculations for example addition and multiplication. See Topic 2.1.

Example

Expand the following:

a) $2(a + b)$
$= 2 \times a + 2 \times b$
$= 2a + 2b$

b) $a(b + c)$
$= a \times b + a \times c$
$= ab + ac$

c) $3(a + 2)$
$= 3 \times a + 3 \times 2$
$= 3a + 6$

d) $a(2a + 3b)$
$= a \times 2a + a \times 3b$
$= 2a^2 + 3ab$

If the term outside the bracket is negative, all the signs of the terms inside the brackets are changed when multiplying out.

Example

Expand the following:

a) $-3(a + b)$
$= -3a - 3b$

b) $-(a - b)$
$= -a + b$

Remember $-(a - b)$ means $-1 \times (a - b)$

c) $-a(a - b)$
$= (-a^2 + ab)$

To simplify expressions, expand the brackets first and then collect like terms.

Example

Expand and simplify the following:

a) $2(a + 3) + 3(a + 1)$

$2a + 6 + 3a + 3$

$= 5a + 9$

b) $5(a + b) - 2(a + 2b)$

$5a + 5b - 2a - 4b$

$= 3a + b$

You will need two sets of cards, each numbered with the digits 2–9. You will also need letter cards (three as, three bs and three cs), three $+$ signs, three $-$ signs and four brackets ( ) ( ). Pick a sign at random, then a number, an open bracket, letter, sign, letter, closed bracket, sign, open bracket, letter, sign, letter and closed bracket. This will make expressions like those shown on this page. On a piece of paper, practise expanding the brackets and then simplifying.

Multiplying out two brackets

Each term in the first bracket is multiplied with each term in the second bracket. A grid method can be used to help when multiplying out two brackets.

> ### Example
>
> Expand and simplify the following:
>
> **a)** $(x + 2)(x + 4)$ or
>
> $x(x + 4) + 2(x + 4)$
> $= x^2 + 4x + 2x + 8$
> $= x^2 + 6x + 8$
>
> | | x | 4 |
> |-------|--------|-------|
> | x | x^2 | $4x$ |
> | 2 | $2x$ | 8 |
>
> $= x^2 + 4x + 2x + 8$
> $= x^2 + 6x + 8$
>
> **b)** $(x - 3)^2$ or
>
> $(x - 3)(x - 3)$
> $= x(x - 3) - 3(x - 3)$
> $= x^2 - 3x - 3x + 9$
> $= x^2 - 6x + 9$
>
> > A common error is to think that $(x - 3)^2$ means $x^2 + (-3)^2 = x^2 + 9$. It does not!
>
> | | x | -3 |
> |-------|--------|--------|
> | x | x^2 | $-3x$ |
> | -3 | $-3x$ | $+9$ |
>
> $= x^2 - 3x - 3x + 9$
> $= x^2 - 6x + 9$
>
> **c)** $(x - a)(x + a) = x(x + a) - a(x + a)$
>
> $= x^2 + ax - ax - a^2 = x^2 - a^2$
>
> > This identity is very important. It is known as the 'difference of two squares'.

Use appropriate cards to make random expressions in the form of $(x + 3)(x - 4)$. Practise expanding and simplifying.

Factorising

Factorising is the reverse of expanding brackets. An expression is put into brackets by taking out common factors.

For example, to factorise $xy + 4y$:

- recognise that y is a factor of each term
- take out the common factor
- the expression is completed inside the bracket, so that the result is equivalent to $xy + 4y$ when multiplied out.

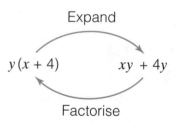

Expand

$y(x + 4)$ $xy + 4y$

Factorise

Here are some more examples:

- $8x - 16 = 8(x - 2)$
- $3x + 18 = 3(x + 6)$
- $5x^2 + x = x(5x + 1)$
- $4x^2 + 8x = 4x(x + 2)$
- $x^3 + 2x^2 + 4x = x(x^2 + 2x + 4)$

Factorising quadratic expressions

You need to be able to factorise quadratic expressions into a pair of linear brackets.

> **Example**
>
> Factorise these.
>
> **a)** $x^2 + 7x + 10$
>
> $x^2 + 7x + 10 = (x \pm a)(x \pm b)$
> $x^2 + 7x + 10 = (x + 5)(x + 2)$
>
> These values multiply to give 10 and add to make 7.
> $5 \times 2 = 10$ and $5 + 2 = 7$
>
> **b)** $x^2 - 4x + 3$
>
> $x^2 - 4x + 3 = (x - 1)(x - 3)$
>
> Since $-1 \times -3 = 3$ and $-1 - 3 = -4$
>
> **c)** $x^2 + 6x - 7$
>
> $x^2 + 6x - 7 = (x + 7)(x - 1)$
>
> Since $7 \times -1 = -7$ and $7 - 1 = 6$
>
> **d)** $x^2 - 16$
>
> $x^2 - 16 = (x + 4)(x - 4)$
>
> This is known as the 'difference of two squares'.

In general, $x^2 - a^2 = (x + a)(x - a)$

Algebraic fractions

You need to be able to add simple algebraic fractions. The same rules that apply to fractions in arithmetic can be used here.

> **Example**
>
> Simplify these.
>
> **a)** $\dfrac{a}{4} + \dfrac{b}{2} = \dfrac{a + 2b}{4}$
>
> Make $\dfrac{b}{2}$ into its equivalent fraction with a denominator of 4, then add the numerators.
>
> **b)** $\dfrac{a}{n} + \dfrac{c}{m} = \dfrac{am}{mn} + \dfrac{cn}{mn} = \dfrac{am + cn}{mn}$
>
> mn is the common denominator.

Working with algebraic fractions is the same as working with numerical fractions. See Topic 1.2.

Algebra is generalised arithmetic so we can use the same methods. See Topic 1.2.

Indices and algebra

The index laws are:

$$a^n \times a^m = a^{n+m}$$

$$a^n \div a^m = a^{n-m}$$

$$(a^n)^m = a^{n \times m}$$

$$a^0 = 1$$

$$a^{-n} = \frac{1}{a^n}$$

$$a^{\frac{1}{n}} = \sqrt[n]{a}$$

$$a^1 = a$$

The laws of indices that apply to numbers also apply to algebra. See Topic 1.1.

Example

Simplify the following:

a) $3x^5 \times 4x^3$

$= 12x^8$

> Note the numbers are multiplied but powers of the same letter are added.

b) $15a^{14} \div 3a^{10}$

$= 5a^4$

c) $(7x^3)^2$

$= 49x^6$

d) x^0

$= 1$

e) $\dfrac{12a^2b^3}{4a^3b}$

$= \dfrac{3b^2}{a}$

$= 3a^{-1}b^2$

f) $2x^{-3}$

$= \dfrac{2}{x^3}$

g) $(2x)^{-3} = \dfrac{1}{(2x)^3}$

$= \dfrac{1}{8x^3}$

> Working out a formula for a pattern of numbers is like finding the nth term of a sequence. See Topic 4.1.

Writing formulae

A formula can be constructed from some information you are given or from a diagram.

Example

A pattern is made up of blue and yellow tiles.

 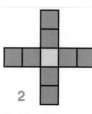 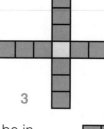 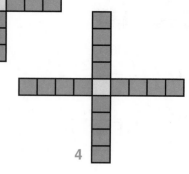

a) How many blue tiles will there be in pattern number 4?

Drawing the diagram, there are 16 blue tiles.

b) Write down the formula for finding the number of tiles (t) in pattern number, n.

Number of tiles

$= 4 \times n + 1$

$t = 4n + 1$

> Make sure there is an = sign in the formula. The 4n is the 4 lots of blue tiles. The + 1 is the yellow tile in the middle.

c) How many tiles will be used in pattern number 12?

In this case, $n = 12$ ← $n = 12$ is substituted into the formula.

$t = 4 \times 12 + 1$

$t = 48 + 1$

$t = 49$

Substituting values into formulae and expressions

Replacing a letter with a number is called **substitution**. When substituting, write out the expression first and then replace the letters with the values given.

Work out the values on your calculator. Use bracket keys where possible and pay attention to the order of operations.

Examples

1. Using $a = 2$, $b = 4.1$, $c = -3$ and $d = 5.25$, find the values of these expressions, giving your answers to 1 decimal place.

a) $\dfrac{a^2 + 2b}{4}$

$= \dfrac{2^2 + 2 \times 4.1}{4}$

$= \dfrac{4 + 8.2}{4}$

$= \dfrac{12.2}{4}$

$= 3.1$ (1 d.p.)

b) $\dfrac{a + 2c}{a - d}$

$= \dfrac{2 + 2 + -3}{2 - 5.25}$

$= \dfrac{-4}{-3.25}$

$= 1.2$ (1 d.p.)

c) $\dfrac{3b^2 (d - 4)}{2a}$

$= \dfrac{3 \times 4.1^2 (5.25 - 4)}{2 \times 2}$

$= \dfrac{50.43 \times (1.25)}{4}$

$= 15.8$ (1 d.p.)

2. The formula $F = \dfrac{9}{5}C + 32$ is used to change temperature in degrees centigrade (C) to temperature in degrees Fahrenheit (F). If $C = 20$, find the value of F.

$F = \dfrac{9}{5}C + 32$

$F = \dfrac{9}{5} \times 20 + 32$

Replace each letter with the given value and calculate carefully.

$F = 68$ degrees

BIDMAS is used in algebra as well as with numbers. See Topic 2.1.

Substituting into formula is used to work out the area and volume of shapes and solids. See Topics 7.2 and 7.3.

Rearranging formulae

The subject of a formula is the letter that appears on its own on one side of the formula. Inverse operations can be used to change the subject.

Examples

1. Make R the subject of the formula $V = IR$

$V = IR$

$\dfrac{V}{I} = R$ or $R = \dfrac{V}{I}$ ← Divide both sides by I.

2. Make l the subject of the formula $T = 2\pi\sqrt{\dfrac{l}{g}}$

$T = 2\pi\sqrt{\dfrac{l}{g}}$

$\dfrac{T}{2\pi} = \sqrt{\dfrac{l}{g}}$ ← Divide both sides by 2π

$\left(\dfrac{T}{2\pi}\right)^2 = \dfrac{l}{g}$ ← Square both sides.

$\left(\dfrac{T}{2\pi}\right)^2 \times g = l$ ← Multiply both sides by g.

$l = \left(\dfrac{T}{2\pi}\right)^2 \times g$ or $l = \dfrac{T^2 g}{4\pi^2}$

Progress Check

1. Simplify these expressions:
   a) $4(x - 2) + 3(x - 1)$
   b) $(n + 1)^2 - 2(n + 2)$

2. Multiply out and simplify:
   a) $(a - b)^2$
   b) $(x - 4)(x + 3)$
   c) $(2a + 3)(2a - 1)$

3. The formula for the perimeter P of a rectangle of length l and width w is $P = 2(l + w)$. Calculate the width of a rectangle if $P = 60$ and $l = 20$

4. Make C the subject of the formula $F = \dfrac{9C}{5} + 32$

5. Simplify $\dfrac{a}{3} + \dfrac{b}{4}$

6. Simplify these:
   a) $2a^4 \times a^6$
   b) $12a^5 \div 2a$
   c) $\dfrac{12a^4b^2}{3ab}$
   d) $(8a^2)^2$
   e) $(4a^{-2})^2$

7. Factorise these:
   a) $x^2 + 8x + 15$
   b) $x^2 + 3x - 10$
   c) $x^2 - 25$

3.2 Equations and inequalities

Linear equations

A **linear** equation has two parts separated by an equals sign. When working out an unknown value in an equation, the balance method is usually used; that is, whatever is done to one side of an equation must be done to the other.

Example

Solve these:

a) $n - 4 = 6$

$n = 6 + 4$ ← Add 4 to both sides.

$n = 10$

b) $5n = 20$

$n = \dfrac{20}{5}$ ← Divide both sides by 5.

$n = 4$

c) $n + 3 = 10$

$n = 10 - 3$ ← Subtract 3 from both sides.

$n = 7$

d) $\dfrac{n}{2} = 8$

$n = 8 \times 2$ ← Multiply both sides by 2.

$n = 16$

Some equations are of the form $ax + b = c$. These equations involve several steps.

Example

Solve the following:

a) $2n - 5 = 11$

$2n = 11 + 5$ ← Add 5 to both sides.

$2n = 16$

$n = \dfrac{16}{2}$ ← Divide both sides by 2.

$n = 8$

b) $\dfrac{n}{4} + 1 = 3$

$\dfrac{n}{4} = 3 - 1$

$\dfrac{n}{4} = 2$

$n = 2 \times 4$ ← Multiply both sides by 4.

$n = 8$

Choose a random value for x between 2 and 9 and make your own equations in the style of those in question 1 in the Progress check on page 71. For example, you might decide that $x = 4$ and your equation might be $5x - 3 = 7$ or $5x - 4 = 6$. Making up your own equations helps you to gain a better understanding of them, especially since you already know the answer!

Some equations are more complicated and have the unknown values on both sides of the equation. These equations are of the form $ax + b = cx + d$. The trick with this type of equation is to get the unknown values together on one side of the equals sign and the numbers on the other side.

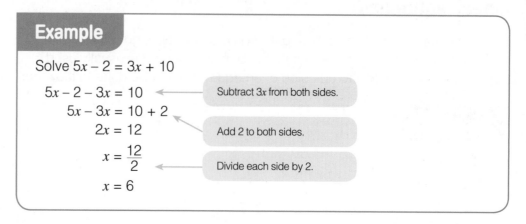

Example

Solve $5x - 2 = 3x + 10$

$$5x - 2 - 3x = 10$$ — Subtract $3x$ from both sides.

$$5x - 3x = 10 + 2$$ — Add 2 to both sides.

$$2x = 12$$

$$x = \frac{12}{2}$$ — Divide each side by 2.

$$x = 6$$

Brackets are often included in more complicated equations. Don't be put off though; it's just the same as solving other equations once the brackets have been multiplied out.

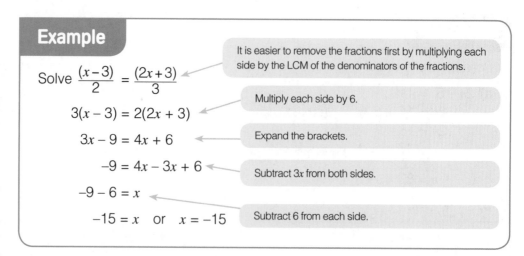

Example

Solve $4(2n + 5) = 3(n - 10)$

$$8n + 20 = 3n - 30$$ — Multiply the brackets out first.

$$8n + 20 - 3n = -30$$

$$5n = -50$$ — Solve as before.

$$n = -\frac{50}{5} = -10$$ — Don't forget the negative sign.

Working on multiples to find the LCM is useful when solving equations with fractions. See Topic 1.1.

Some of the previous equations had solutions involving fractions. Some equations have fractions in them.

Example

Solve $\frac{(x - 3)}{2} = \frac{(2x + 3)}{3}$ — It is easier to remove the fractions first by multiplying each side by the LCM of the denominators of the fractions.

$$3(x - 3) = 2(2x + 3)$$ — Multiply each side by 6.

$$3x - 9 = 4x + 6$$ — Expand the brackets.

$$-9 = 4x - 3x + 6$$ — Subtract $3x$ from both sides.

$$-9 - 6 = x$$

$$-15 = x \quad \text{or} \quad x = -15$$ — Subtract 6 from each side.

Using equations to solve problems

When setting up equations, the information you are given will include an unknown quantity. State the letter you decide to use to represent this quantity.

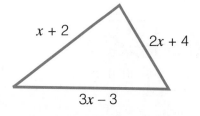

Examples

1. The lengths of the sides of a triangle are given in the diagram below.

   a) Write down an expression for the perimeter of the triangle.

   Perimeter = $(x + 2) + (3x - 3) + (2x + 4)$

   = $6x + 3$

   > The perimeter is found by adding the three lengths.

   b) If the perimeter of the triangle is 39cm, form an equation and solve it to find the length of each side.

   $$6x + 3 = 39$$
   $$6x = 39 - 3$$
   $$6x = 36$$
   $$x = \frac{36}{6}$$
   $$= 6$$

   The sides are: $x + 2 = 8$cm
   $$2x + 4 = 16\text{cm}$$
   $$3x - 3 = 15\text{cm}$$

2. In this triangular arithmagon, what could the numbers x, y and z be? (In an arithmagon, the number in a square is the sum of the numbers in the two circles on either side of it.)

   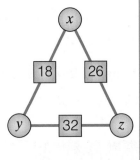

   Let x stand for the number in the top circle. Two expressions can be formed for the other two numbers in the circles. Then form an equation to find x.

   Since $x + y = 18$ rearranging $y = 18 - x$
   and $x + z = 26$ rearranging $z = 26 - x$
   and $y + z = 32$ this equation becomes $(18 - x) + (26 - x) = 32$
   $$44 - 2x = 32$$
   $$44 - 32 = 2x$$
   $$12 = 2x$$
   $$x = 6$$

   So $x = 6$, $y = 12$, $z = 20$

> Setting up equations to solve problems is used in trigonometry and solving area and volume problems. See Topics 5.4, 7.2 and 7.3.

> Think of some real-life problems of your own and make equations for them. Always give the unknown a letter. For instance, you might try to make a problem out of family members' ages. Mum is 25 years older than me. Together our ages total 39. How old am I? Your age becomes x so your equation is $x + 25 = 39$.

Simultaneous linear equations

Two equations both with two unknowns are called **simultaneous linear equations**. They can be solved in several ways. Solving equations simultaneously involves finding values for the letters that will make both equations work.

Elimination method

If the coefficient of one of the letters is the same in both equations then that letter can be eliminated by adding or subtracting the equations.

A **coefficient** is a number in front of a letter. For example, 2 is the coefficient of $2n$.

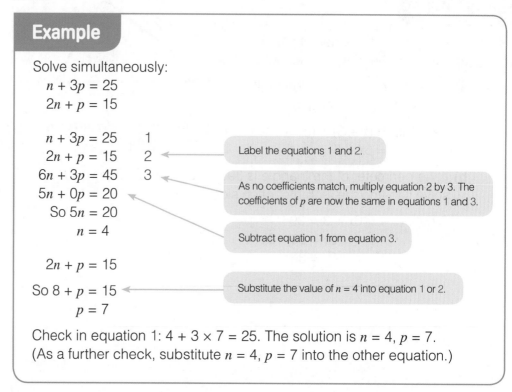

Example

Solve simultaneously:
$$n + 3p = 25$$
$$2n + p = 15$$

$n + 3p = 25$	1
$2n + p = 15$	2
$6n + 3p = 45$	3
$5n + 0p = 20$	
So $5n = 20$	
$n = 4$	

Label the equations 1 and 2.

As no coefficients match, multiply equation 2 by 3. The coefficients of p are now the same in equations 1 and 3.

Subtract equation 1 from equation 3.

$$2n + p = 15$$
So $8 + p = 15$
$$p = 7$$

Substitute the value of $n = 4$ into equation 1 or 2.

Check in equation 1: $4 + 3 \times 7 = 25$. The solution is $n = 4$, $p = 7$.
(As a further check, substitute $n = 4$, $p = 7$ into the other equation.)

To eliminate terms with **opposite** signs, **add** the two equations. To eliminate terms with **the same** signs, **subtract** the two equations.

Substitution method

Simultaneous equations can also be solved by writing one of the equations in the form '$x = ...$' or '$y = ...$'. This is called substitution.

Example

Solve
$$2x - y = 2 \quad\quad 1$$
$$3x + 2y = 17 \quad\quad 2$$

Rearrange equation 1 to give $2x - 2 = y$

Substitute $2x - 2 = y$ into equation 2.

$$3x + 2(2x - 2) = 17$$
$$3x + 4x - 4 = 17$$
$$7x = 17 + 4$$
$$7x = 21$$
$$x = 3$$

Work out the value of x from this equation.

$$y = 2x - 2$$
$$y = 2 \times 3 - 2$$
$$y = 4$$
$$x = 3, y = 4$$

Substitute 3 for x in the first equation.

Remember to check your solution.

Have a go at making up your own simultaneous equations. Choose random values for x and y. Make up two equations and solve them using elimination and substitution.

Graphical method

The point at which two straight-line graphs intersect also gives the simultaneous solution of their equations.

Example

Solve the simultaneous equations $y = 2x - 3$, $y - x = 1$ by a graphical method.

Draw the two graphs:

$y = 2x - 3$ if $x = 0$, $y = -3$

            if $y = 0$, $x = \frac{3}{2}$

$y - x = 1$ if $x = 0$, $y = 1$

            if $y = 0$, $x = -1$

At the point of intersection $x = 4$ and $y = 5$.

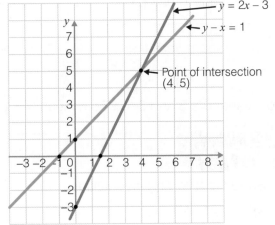

Point of intersection (4, 5)

Solving cubic equations by trial and improvement

In trial and improvement, successive approximations are made in order to get closer to the correct value.

> When solving equations by trial and improvement, we only get an estimate, which is usually rounded to 1 or 2 decimal places. See Topic 2.2.

Example

The equation $x^3 - 5x = 10$ has a solution between 2 and 3. Find this solution to two decimal places.

Drawing a table may help and then substitute different values of x into $x^3 - 5x$.

x	$x^3 - 5x$	Comment
2.5	3.125	Too small
2.8	7.952	Too small
2.9	9.889	Too small
2.95	10.922375	Too big
2.94	10.712184	Too big
2.91	10.092171	Too big

At this stage the solution is trapped between 2.90 and 2.91

Checking the middle value $x = 2.905$ gives $x^3 - 5x = 9.990\,36...$ which is too small. Because $x = 2.905$ is too small, the solution is 2.91 correct to two decimal places.

2.90	2.905	2.91
(Too small)	(Too small)	(Too big)

Inequalities

Inequalities are solved in a similar way to equations. Multiplying and dividing by negative numbers changes the direction of the sign. For example, if $-x \geqslant 5$ then $x \leqslant -5$.

The four inequality symbols are:

$>$ means 'greater than'	$\geqslant$ means 'greater than or equal to'
$<$ means 'less than'	$\leqslant$ 'means less than or equal to'.

Notice that $x > 3$ and $3 < x$ both mean 'x is greater than 3'.

Example

Solve the following inequalities.

a) $4x - 2 < 6$

$4x < 6 + 2$ ⟵ Add 2 to both sides.

$4x < 8$

$x < \dfrac{8}{4}$ ⟵ Divide both sides by 4.

$x < 2$

The solution of the inequality can be represented on a number line:

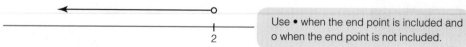

Use ● when the end point is included and o when the end point is not included.

b) $-5 < 3x + 1 \leqslant 13$

$-6 < 3x \leqslant 12$ ⟵ Subtract 1 from each side.

$-2 < x \leqslant 4$ ⟵ Divide each side by 3.

The integer values that satisfy the above inequality are $-1, 0, 1, 2, 3, 4$

Graphs of inequalities

The graph of the equation $y = 3$ is a straight line, whereas the graph of the inequality $y < 3$ is a region that has the line $y = 3$ as its boundary.

This is how to show the region for given inequalities:
- Draw the boundary lines first.
- For strict inequalities $>$ and $<$, the boundary line is not included and is shown as a dotted line.
- It is often easier with several inequalities to shade out the unwanted regions, so that the solution is shown unshaded.

For example, the diagram shows unshaded the region $x > 1$, $x + y \leq 4$, $y \geq 0$.

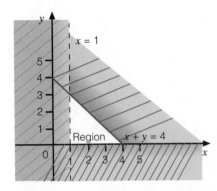

1. Solve the following equations.
   a) $5x - 2 = 12$
   b) $4x + 2 = 18$
   c) $5x + 3 = 2x + 9$
   d) $6x - 1 = 15 + 2x$
   e) $2(x - 1) = 6(2x + 2)$
   f) $3(n + 1) + 4(n + 2) = 39$

2.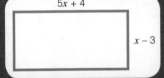

   a) The perimeter of this rectangle is 74cm. Write down an equation for the perimeter.
   b) Solve the equation to find the length and width of the rectangle.

3. The equation $y^3 + y = 40$ has a solution between 3 and 4. Find this solution to 1 d.p. by using a method of trial and improvement. 🖩

4. Solve the following pair of simultaneous equations.
   $3a - 5b = 1$
   $2a + 3b = 7$

5. Solve the following inequalities.
   a) $2x - 3 < 9$
   b) $5x + 1 \geq 21$
   c) $1 \leq 3x - 2 \leq 7$

Progress Check

Worked questions

1. Solve the equation:

$$(2x - 4)^2 - 2x(2x - 2) = 26 - 7x \qquad \textit{(2 marks)}$$

First of all, expand the brackets.

$(2x - 4)(2x - 4) - 2x(2x - 2) = 26 - 7x$
$4x^2 - 8x - 8x + 16 - 4x^2 + 4x = 26 - 7x$

Now gather the like terms.

$4x^2 - 4x^2 = 0$
$-8x - 8x + 4x = -12x$
$-12x + 16 = 26 - 7x$

Gather the unknowns on the left hand side and the ordinary numbers on the right. Add $7x$ to each side.

$-12x + 16 \, (+7x) = 26 - 7x \, (+7x)$
$-5x + 16 = 26$

Take 16 from each side.

$-5x + 16 \, (-16) = 26 \, (-16)$
$-5x = 10$

The unknown should always be positive. To make the number positive and to isolate x, divide both sides by –5.

$x = -2$

Many problems can be solved by giving the unknown quantity a letter, thereby creating one linear equation. Often, as in this question, there will be two or more unknowns but they will have a relationship which enables them all to be expressed in terms of just one of them.

2. Last week, Junaid spent three hours more on his exam revision than Jenny. Jenny spent 4 hours more than Jon. In total they all spent 65 hours on exam revision.

How much did each student spend individually? \qquad \textit{(2 marks)}

Give the amount of time that Junaid spent the letter x.

Junaid = x
So Jenny = $x - 3$
And Jon = $x - 7$

Gather the like terms.

$x + x - 3 + x - 7 = 65$

$3x - 10 = 65$

Isolate the unknown by adding 10 to each side.

$3x = 75$

Divide both sides by 3.

$x = 25$

It is always a good idea to check that your final amounts equal what you started with:
$25 + 22 + 18 = 65$

Junaid spent 25 hours, Jenny spent 22 hours and Jon spent 18 hours.

If you have the same number of one of the unknowns, elimination is easy. You just have to be careful with your signs. It would be so easy to look at this and think you just need to subtract to eliminate the ys but $+4$ minus $-4 = 4 + 4 = 8$. These equations must be added to eliminate the ys. $+4 + -4 = 4 - 4 = 0$

3. Solve these simultaneous equations by elimination.

a) $6x + 4y = 20$
$9x - 4y = 10$ \qquad \textit{(3 marks)}

$6x + 4y = 20 \qquad$ equation 1
$9x - 4y = 10 \qquad$ equation 2

$15x = 30$
$x = 2$

$$(6 \times 2) + 4y = 20$$

$$12 + 4y = 20$$

$$4y = 20 - 12$$
$$4y = 8$$
$$y = 2$$

The values are $x = 2$ and $y = 2$

b) $3x - 2y = 12$
$2x + 5y = 27$

(3 marks)

$3x - 2y = 12$ equation 1
$2x + 5y = 27$ equation 2

$3x - 2y = 12$ x2

$2x + 5y = 27$ x3

$6x - 4y = 24$
$6x + 15y = 81$

$15y - -4y = 15y + 4y = 19y$

$19y = 57$
$y = 3$

$3x - 2y = 12$

$3x - 6 = 12$
$3x = 18$
$x = 6$

The values are $x = 6$ and $y = 3$

4. Solve these simultaneous equations by substitution.

$3x - y = 3$
$7x + y = 17$

$3x - y = 3$ equation 1
$7x + y = 17$ equation 2

$y = 3x - 3$

$7x + (3x - 3) = 17$
$10x - 3 = 17$
$10x = 17 + 3$
$x = 2$

Put this value into equation 1.

The unknown value needs to be on its own, preferably on the left side of the equation. To do this, you must subtract the 12.

Whatever you do to one side you do to the other! Divide both sides by 4.

It is always a good idea to check that the values work in the other equation:
$9x - 4y = 10$
$18 - 8 = 10$

Here, you cannot eliminate immediately by adding or subtracting. You have to make one of the values equal by multiplying. It doesn't matter whether you choose x or y. Work by the principle that it is better to keep your numbers as low as possible. So choose x.

Remember to multiply all the elements of each equation so that you do not alter the original values.

Now take equation 1 from equation 2. Again, take care with your signs.

Divide both sides by 19.

Put this value into equation 1.

Add 6 to both sides.

Again, check your values work in equation 2:
$2x + 5y = 27$
$12 + 15 = 27$

Make y the subject of one of the equations. Again, choose the one which will keep your numbers low. Here, it is equation 1.

Substitute this expression for the y value in equation 2.

Put this value into equation 1.

$$3x - y = 3$$
$$6 - y = 3$$
$$y = 6 - 3$$
$$y = 3$$

Check your values in equation 2:
$14 + 3 = 17$

The values are $x = 2$ and $y = 3$

5. George bought four portions of chips and three fish for £5.55. Ali bought two portions of chips and four fish for £4.90.
How much did Harriet pay for one portion of chips and two fish?  *(4 marks)*

There are two unknown values: the price of the chips and the price of the fish. Give each a symbol; chips = c and fish = f and then make two simultaneous equations.

$$4c + 3f = £5.55 \quad \text{equation 1}$$
$$2c + 4f = £4.90 \quad \text{equation 2}$$

You can't eliminate by adding or subtracting, so make equivalent equations.

$$4c + 3f = £5.55 \quad × 2 = 8c + 6f = £11.10$$
$$2c + 4f = £4.90 \quad × 4 = 8c + 16f = £19.60$$

Subtract equation 1 from equation 2.

$$10f = £8.50$$
$$f = 85p$$

Put this value into equation 1.

$$4c + £2.55 = £5.55$$
$$4c = £5.55 - £2.55$$
$$4c = £3.00$$
$$c = 75p$$

Harriet paid £2.45 for one portion of chips and two fish.

6. Find all the integer values of x which satisfy $-3 \leqslant 3x < 18$. Show all the values of x on a number line. *(2 marks)*

Divide the inequality by 3.

$\leqslant$ means less than or equal to which means that -1 is a possible value.

$$-3 \leqslant 3x < 18$$
$$-1 \leqslant x < 6$$

The values of x are $-1, 0, 1, 2, 3, 4$ and 5.

$$-3 \quad -2 \quad -1 \quad 0 \quad 1 \quad 2 \quad 3 \quad 4 \quad 5 \quad 6$$

7. Solve $4x < 6x + 8$ *(2 marks)*

You need the xs on the left so subtract $6x$ from each side.

x should be positive. Multiply both sides by -1. Multiplying by a negative number reverses the sign.

$$4x - 6x < 8$$
$$-2x < 8$$
$$-x < 4$$

$$x > -4$$

Practice questions

1. If $a = -2$, $b = 3$, $c = 4$ and $d = 5$, find the value of:

   a) $3b^2 - 4c$ *(1 mark)*

   b) $bcd + \dfrac{3c}{ba^2}$ *(1 mark)*

2. **a)** The length of each side of a regular hexagon is $(3x + 1)$ cm.

   Write, and then expand, an expression for the perimeter of the hexagon. *(2 marks)*

   b) A triangle has sides of $y + 2$, $2y + 4$ and $4y - 3$.

   Write, and then simplify, an expression for the perimeter of the triangle. *(2 marks)*

   c) At the Coliseum Theatre, tickets for the stalls cost £15, the circle £20 and the balcony £10. There are s number of seats in the stalls, c number of seats in the circle and b number of seats in the balcony.

   Write an expression to illustrate how much money the box office will take when the theatre is full. *(1 mark)*

3. Simplify these expressions:

   a) $ab^2 + 3a^2b - 2a^2b + 3b^2a - 2ab^2$ *(1 mark)*

   b) $7(2s + t) - 6s + 5(3t - s)$ *(1 mark)*

   c) $(3 + 4x)(2y - 2)$ *(1 mark)*

   d) $(3 - p)(q + 1) + 4(p + pq + 2q) + 3qp$ *(1 mark)*

4. Rearrange the following to make p the subject:

   a) $b = 2p$ *(1 mark)*

   b) $z = 3p + 4$ *(1 mark)*

   c) $q = \dfrac{2s}{p}$ *(1 mark)*

   d) $r = 3p^2q$ *(1 mark)*

5. Simplify the following:

   a) $-3xy \times -2xy^2$ *(1 mark)*

   b) $4x^4 \times 2x^3$ *(1 mark)*

   c) $\dfrac{30y^6}{5y^2}$ *(1 mark)*

6. Expand the following expressions:

   a) $(x + 4)(x + 1)$ *(1 mark)*

   b) $(x - 5)(x + 2)$ *(1 mark)*

   c) $(x - 3)(x - 1)$ *(1 mark)*

   d) $(x + 3)^2$ *(1 mark)*

   e) $(x - 4)^2$ *(1 mark)*

7. Factorise: *(1 mark)*

    a) $8x^2 + 12x$ *(1 mark)*

    b) $6ab + 9ab^2 - 3a^2b$ *(1 mark)*

    c) $r^2 - 64$ *(1 mark)*

    d) $y^2 + 7y + 6$ *(1 mark)*

8. Solve these linear equations.

    a) $3p + 7 = 4$ *(1 mark)*

    b) $\frac{q}{8} = 4$ *(1 mark)*

    c) $4x - 6 = x + 9$ *(1 mark)*

    d) $2(2x + 4) = 6(3x - 1)$ *(1 mark)*

9. Saima thinks of a number. She multiplies it by 3, divides it by 7 and adds 5. The result is 3 less than the number she started with.

    a) Using this information set up an equation in x to show the number Saima thought of. *(1 mark)*

    b) Solve the equation. *(1 mark)*

10. In a spelling test, Mary scored 5 more than James. James scored 4 less than Alex. In total they scored 30.

    a) Make an equation in x to show this information. *(1 mark)*

    b) Solve the equation to find out each student's score in the spelling test. *(2 marks)*

11.   **a)** Amy makes a return journey to work of $2x$km on four days per week. On Fridays she works from home and at the weekend she travels on average a total of 15km. In all, she travels 71km per week.

       i) Make an equation in x to show this information. *(1 mark)*

       ii) Solve the equation to find the distance from Amy's home to her place of work. *(1 mark)*

    b) In one week Amy jogs 35km further than her friend Fred. Altogether Amy and Fred jog 67km. *(1 mark)*

       i) Use p to represent how far Fred jogs and make an equation for Amy's distance. *(1 mark)*

       ii) Solve the equation to find out how far Fred jogs each week. *(1 mark)*

12. Simplify:

    a) $\dfrac{x + 3}{3} + \dfrac{x - 2}{3}$ *(1 mark)*

    b) $\dfrac{x + 1}{3} - \dfrac{2x}{7}$ *(1 mark)*

    c) $\dfrac{3x}{2y} \times \dfrac{y}{2}$ *(2 marks)*

    d) $\dfrac{y}{2x} \div \dfrac{3}{4x}$ *(2 marks)*

13. Solve these simultaneous equations.

$4x + 2y = 46$

$3x - 2y = 10$ *(3 marks)*

14. Rearrange this formula to make y the subject.

$y + 4x = 6$ *(1 mark)*

15. Solve these simultaneous equations by substitution.

$y + 4x = 6$

$3y - 2x = 4$ *(3 marks)*

16. Joseph buys four adult tickets and two children's tickets for the circus and pays £44. Farrah buys three adult tickets and three children's tickets and pays £41.25.

Find the price of: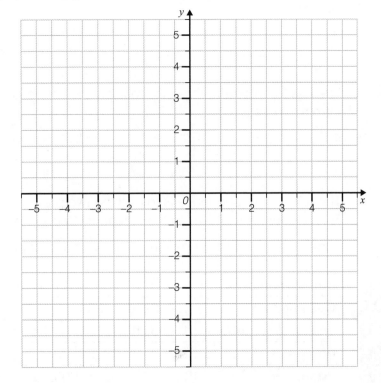

a) one adult ticket. *(2 marks)*

b) one child's ticket. *(1 mark)*

c) two adult tickets and one child's ticket. *(1 mark)*

17. Use trial and improvement to solve these equations.

a) $x^2 + 4x = 26$

The answer lies between 3 and 4. Give your answer to 1 decimal place. *(3 marks)*

b) $x^3 - 2x = 52$

The answer lies between 3 and 4. Give your answer to 2 decimal places. *(4 marks)*

18. a) Find all the integer values of x where $-4 < x \leqslant 2$ *(1 mark)*

b) Show these values on a number line. *(1 mark)*

c) Show on a graph the region which satisfies $-4 < x \leqslant 2$ *(3 marks)*

19. Find the value of x.

$$\sqrt{\frac{4 + 3x - 6}{x + 2}} = 2$$ *(2 marks)*

4 Sequences, functions and graphs

Learning Summary

After studying this chapter you should be able to:
- generate terms of a sequence and write an expression to describe the nth term of a sequence
- find the next term and the nth term of a quadratic sequence
- draw, interpret and identify graphs of a variety of functions
- construct functions arising from real-life problems.

4.1 Sequences and functions

Sequences

A **sequence** is a list of numbers. There is usually a relationship between the numbers. Each value in the list is called a **term**.

There are lots of different number patterns. When finding a missing number in the number pattern, it is sensible to look for the 'term-to-term' rule.

A 'term-to-term' rule tells you the next term by looking at the difference between consecutive terms.

For example:

2   6   18   54...

    ×3   ×3   ×3

The rule for this sequence is to multiply the previous term by 3. The term-to-term rule here is ×3.

1   1   2   3   5   8

    1 + 1   1 + 2   2 + 3   3 + 5

The rule for this sequence is to add the two previous numbers each time. This sequence is known as the Fibonacci Sequence.

Some common number patterns you need to recognise are shown.

Square numbers	1, 4, 9, 16, 25, 36, 49, 64, 81, 100, ...
Cube numbers	1, 8, 27, 64, 125, ...
Triangular numbers	1, 3, 6, 10, 15, 21, 28, 36, 45, 55, 66, 78, 91, ...
Powers of 2	2, 4, 8, 16, 32, 64, ...
Powers of 10	10, 100, 1000, 10000, 100000, ...

Practise writing down these common number patterns. Make square, cube and triangular patterns with building blocks.

Finding the nth term of a linear sequence

The nth term is sometimes denoted by T(n), where:
- T(1) = first term
- T(2) = second term
- T(n) = n^{th} term, etc.

For a linear sequence, the nth term takes the form T(n) = $an + b$

If you plot the term numbers (1, 2, 3, ...) against the terms of a linear sequence, the graph is a straight line.

A straight line graph represents a linear sequence. See Topic 4.1.

Example

1. If T(n) = $3n - 1$, write down the first five terms of the sequence.

   $n = 1$ T(1) = $3 \times 1 - 1 = 2$
   $n = 2$ T(2) = $3 \times 2 - 1 = 5$
   $n = 3$ T(3) = $3 \times 3 - 1 = 8$
   $n = 4$ T(4) = $3 \times 4 - 1 = 11$
   $n = 5$ T(5) = $3 \times 5 - 1 = 14$

   The first five terms are 2, 5, 8, 11, 14.

2. Find the nth term of this sequence: 4, 7, 10, 13, 16, ...

   Look at the difference between the terms. If the difference is the same number this is the value of a or the multiple.

   Adjust the rule by adding or subtracting a value, which is b.

Term	1	2	3	4	5.....n
Sequence	4	7	10	13	16

   1st difference   3   3   3   3

   The multiple is 3, i.e. $3n$.

   If $n = 1$ then $3 \times 1 = 3$ but we need 4, so we adjust by adding 1.
   n^{th} term T(n) = $3n + 1$.

   Check with the second term: T(2) = $3 \times 2 + 1 = 7$

Make up your own sequences using a constant difference. Write them down. Use your calculator to make sequences with larger numbers.

Finding the nth term of a quadratic sequence

For a quadratic sequence, the first differences are not constant but the second differences are.

The nth term T(n) takes the form T(n) = $an^2 + bn + c$, where b and c may be zero.

If you plot the term numbers against the terms of a quadratic sequence, the graph is a parabola. See Topic 4.2.

Example

1. Find the nth term of this sequence.

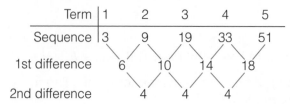

Term	1	2	3	4	5
Sequence	3	9	19	33	51

1st difference: 6, 10, 14, 18

2nd difference: 4, 4, 4

Since the second differences are the same then the rule for the nth term is quadratic.

The coefficient of n^2 is (the second difference) $\div$ 2 i.e. $4 \div 2 = 2$
Adjusting as with linear sequences gives $2n^2 + 1$.

2. Find the first five terms of the sequence $T(n) = n^2 + 2n - 1$.

$T(1) = 1^2 + 2 \times 1 - 1 = 2$
$T(2) = 2^2 + 2 \times 2 - 1 = 7$
$T(3) = 3^2 + 2 \times 3 - 1 = 14$
$T(4) = 4^2 + 2 \times 4 - 1 = 23$
$T(5) = 5^2 + 2 \times 5 - 1 = 34$

Fraction sequences

When finding the nth term of fraction sequences, it is usually better to look at the numerator and denominator separately.

Example

Find the nth term of the following fraction sequences:

a) $\frac{1}{2}, \frac{2}{3}, \frac{3}{4}, \frac{4}{5} \cdots$

$T(n) = \frac{n}{n+1}$

b) $\frac{1}{2}, \frac{1}{4}, \frac{1}{6}, \frac{1}{8} \cdots$

$T(n) = \frac{1}{2n}$

c) $\frac{1}{2}, \frac{1}{4}, \frac{1}{8}, \frac{1}{16} \cdots$

$T(n) = \frac{1}{2^n} = 2^{-n}$

Generating sequences from practical examples

The nth term can usually be found by looking at the practical context from which it arose. For example:

Number of squares	1	2	3	4	...
Number of matches	4	7	10	13	...

In the nth arrangement $T(n) = 3n + 1$.

This can be justified by looking at the structure of the shape: each square needs three matches plus an extra one for the first square.

For n squares, $3n$ matches are needed plus 1 for the first square.

Use spent matches to make similar patterns to those shown in the text. Make a table and work out how many matches would be needed for the nth pattern.

Plotting the values in the table i.e. the number of squares against the number of matches will give a straight line graph, since it is a linear function. See Topic 4.2.

Function machines and mapping

Function machines are useful when finding a relationship between two variables. For example:

Input (x) ⟶ ×2 ⟶ +1 ⟶ Output (y)

When numbers are fed into this machine they are first multiplied by 2 and 1 is then added.

If 1 is fed in, 3 comes out $(1 \times 2 + 1 = 3)$
If 2 is fed in, 5 comes out $(2 \times 2 + 1 = 5)$, etc.

This transformation can be illustrated with a **mapping** diagram like this:

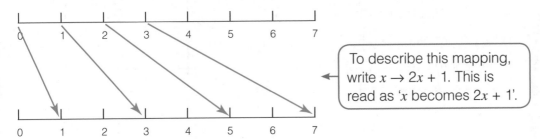

To describe this mapping, write $x \rightarrow 2x + 1$. This is read as 'x becomes $2x + 1$'.

The mapping $x \rightarrow x$ is called the **identity function** because it maps any number onto itself.

---

1. Here is a pattern made up of regular hexagons with sides 1cm.
   The table shows the pattern number (n) and the perimeter of each shape.

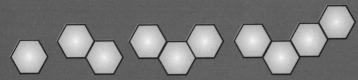

   a) Complete the table.

Pattern number (n)	1	2	3	4	5	6
Perimeter (cm)	6		14		22	

Progress Check

   b) What would be the perimeter for the nth pattern?
   c) What would be the perimeter for pattern number 50?
2. Continue the following sequences for the next two terms:
   a) 10, 13, 16, 19, ...   **b)** 1, $\frac{1}{2}$, $\frac{1}{4}$, $\frac{1}{8}$ ...   **c)** 1, –3, 9, –27, ...
3. Write down the nth term, T(n), of these sequences:
   a) 5, 7, 9, 11, ...   **b)** 7, 10, 13, 16, ...   **c)** 2, 8, 18, 32, ...
4. Write down the first four terms of these sequences:
   a) T(n) = $n^2 + 1$ **b)** T(n) = $10 - 2n$ **c)** T(n) = $n^2 + 4n - 6$

4.2 Graphs of functions

Coordinates

> We use both positive and negative numbers when reading coordinates. See Topic 1.1.

Coordinates are used to locate the position of a point. When reading coordinates read across first, then up or down.

Coordinates are always written in brackets and with a comma in between, e.g. (2, 4). The horizontal axis is the x-axis. The vertical axis is the y-axis.

For example:

A has coordinates (2, 4)

B has coordinates (−1, 3)

C has coordinates (−2, −3)

D has coordinates (3, −1)

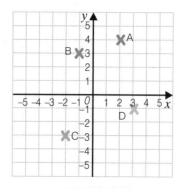

> Using 5mm squared graph paper, practise plotting a variety of coordinates. Work out pairs which make horizontal and vertical lines. Look at the gradients particular pairs make. Plot points which make a variety of 2-D shapes.

Straight-line graphs

Graphs of the form $x = b$ and $y = a$

$y = a$ is a **horizontal** line, with every y-coordinate equal to a.

$x = b$ is a **vertical** line, with every x-coordinate equal to b.

> If you have access to a large outdoor space, you could chalk out your own grid with x and y axes and walk to different sets of coordinates. Remember to go along the horizontal before you walk up or down the vertical.

Example

a) Draw the line $y = 3$.

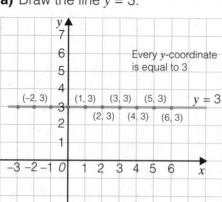

b) Draw the line $x = 2$.

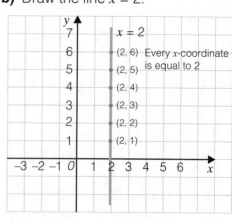

Coordinates are used to draw graphs. Before a graph can be drawn the coordinates have to be worked out.

To work out the coordinates for the graph, you can do either of the following:

- Draw up a table as shown in the examples.
- Use a function machine and mapping diagrams.

Graphs of the form $y = mx + c$

Graphs of the form $y = mx + c$ are straight-line (linear) graphs. To draw a straight-line graph you need at least three sets of coordinates.

Example

Draw the graph of $y = 2x + 1$.

Choose some values of x, e.g. -2, 0, 2.
Replace x in the function with the given values.

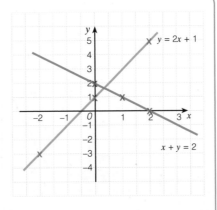

	$x \to 2x + 1$	Coordinates
-2	$\to (2 \times -2 + 1) = -3$	$(-2, -3)$
0	$\to (2 \times 0 + 1) = 1$	$(0, 1)$
2	$\to (2 \times 2 + 1) = 5$	$(2, 5)$

Alternatively, a table of values can be used:

x	-2	0	2
y	-3	1	5

Plot the coordinates and join up the points with a straight line. Label the graph.

For the graph $x + y = 2$, choose some values of x and work out the corresponding values of y:

If $x = 0$, $y = 2$; if $x = 1$, $y = 1$; if $x = 2$, $y = 0$

> Make up your own simple functions, e.g. $y = x + 4$, $y = 9x$, $y = 8 - x$, $y = x - 3$. Draw a table like the one opposite but ranging from -3 to $+3$. Work out the value of y when x equals each of the values.

Linear functions can be rearranged to give y in terms of x and the coordinates can be worked out as normal.

For example, if drawing the graphs of:
- $y - 2x + 2 = 0 \to$ rearrange to give $y = 2x - 2$
- $2y + 3x = 6 \to$ rearrange to give $2y = 6 - 3x$, i.e. $y = 3 - \frac{3}{2}x$
- $x + y = 2 \to$ rearrange to give $y = 2 - x$ or draw directly $x + y = 2$ (see above).

The general equation of a straight-line graph is $y = mx + c$. m is the **gradient** (steepness) of the line:
- As m increases, the line gets steeper.
- If m is positive, the line slopes upwards (from left to right).
- If m is negative, the line slopes downwards (from left to right).

For example:

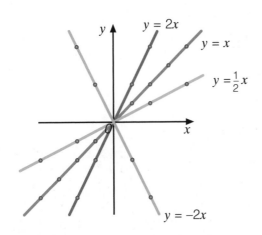

Parallel lines have the same gradient. Each of the straight lines below has a gradient of 2.

c is the **intercept** on the y-axis, that is where the graph cuts the y-axis.

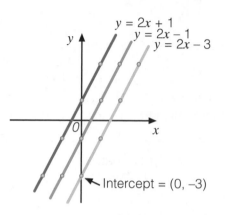

Finding the gradient of a straight line

To find the gradient of a straight line, choose any two points on the line.

$$\text{Gradient} = \frac{\text{change in } y}{\text{change in } x}$$

For example, to find the gradient of this straight line choose two points on the line. Find the change in y (height) and the change in x (base).

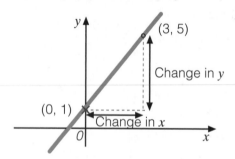

$$\text{Gradient} = \frac{\text{change in } y}{\text{change in } x} \text{ or } \frac{\text{height}}{\text{base}} = \frac{4}{3} = 1\frac{1}{3}$$

Decide if the gradient is positive or negative. In this case, it is positive.

Example

Write down the gradient and intercept for each of these straight-line graphs:

a) $y = 4x - 3$
Gradient = 4
Intercept = (0, –3)

b) $y = 6 - 2x$
Gradient = –2
Intercept = (0, 6)

c) $2y + 10 = 4x$ ← This equation needs to be rearranged into the form $y = mx + c$.
Gradient = 2
Intercept = (0, –5)

Using 5mm squared graph paper, draw different straight lines and practise working out the gradients of them. You should draw lines which slope in different directions to identify negative and positive gradients. Identify gradients which produce parallel lines.

Piece-wise linear function

The graph shown below is called piece-wise because it consists of two or more pieces:

- $y = 2x + 3$ is in the interval $x = -3$ to $x = 1$
- $y = 5$ is in the interval $x = 1$ to $x = 5$.

The graph spans the interval from $x = -3$ to $x = 5$. The endpoints are not included so the point where each line starts and stops is shown by an open circle.

> We use the same notation with the open circles as we do when representing inequalities on a number line. See Topic 3.2.

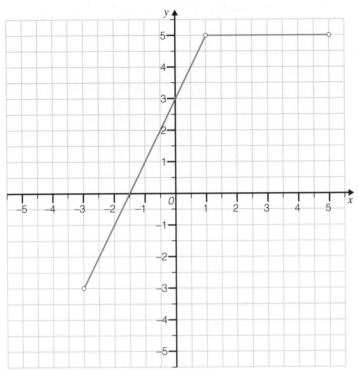

We can see that the gradient of the function is not constant throughout the graph. The gradient is 2 in the first piece and 0 in the second piece. However, at the point of intersection, when we substitute 1 in for x into $y = 2x + 3$, we get $y = 5$ for both functions, so the lines share the point $(1, 5)$.

Graphs that are not straight lines

Quadratic graphs

Quadratic graphs are curved graphs of the form $y = ax^2 + bx + c$ where $a \neq 0$. $\neq$ means is not equal to.

If the number in front of x^2 is positive, the curve looks like this: 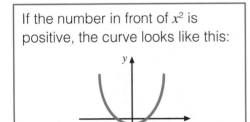	If the number in front of x^2 is negative, the curve looks like this: 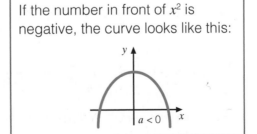

Example

Draw the graph of $y = x^2 - x - 6$ using values of x from −2 to 3. Use the graph to find the value of x when $y = -3$.

Make a table of values:

x	−2	−1	0	1	2	3	0.5
y	0	−4	−6	−6	−4	0	−6.25

Replace x in the equation with each value, i.e.
when $x = -2$, $y = (-2)^2 - (-2) - 6$,
$$y = 0$$

The table represents the coordinates of the graph, which can now be plotted. Join the points with a smooth curve and label the graph.

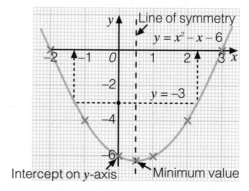

The minimum value is when $x = 0.5$ and $y = -6.25$
The line of symmetry is at $x = 0.5$.
The curve cuts the y-axis at $(0, -6)$, i.e. $(0, c)$.
To find the value of x when $y = -3$, read across from $y = -3$ to the graph then read up to the x-axis.
$x = 2.3$ and $x = -1.3$. These are the approximate solutions of the equation $x^2 - x - 6 = -3$.

If you are asked to draw the graph of $y = 2x^2$, remember that this means $y = 2 \times x^2$, i.e. square x first and then multiply by 2.

Cubic graphs

When drawing the graph of $y = x^3$, it is important to remember that $x^3 = x \times x \times x$

Example

Draw the graph of $y = x^3$

Work out the y-coordinate for each point.
Replace x in the equation with each value.

x	−3	−2	−1	0	1	2	3
y	−27	−8	−1	0	1	8	27

Plot the x and y-coordinates from the table above.

Reciprocal graphs

The graph of the equation $y = \frac{a}{x}$ takes one of two forms, depending on the value of a.

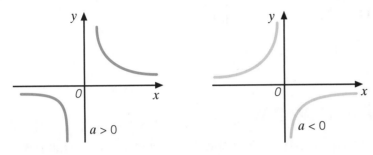

Exponential graphs

The graph of the form $y = a^x$ is known as an exponential graph. The graph can be used to show exponential growth, for example compound interest or the growth of bacteria.

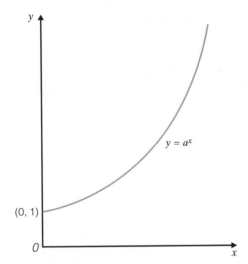

1. The graph of $y = x - 1$ is shown opposite.
   Draw the following graphs on the same axes.
   a) $y = 2x$ **b)** $y = 4x$
   c) What do you notice about the graphs
   $y = 2x$ and $y = 4x$?
   d) Without working out any coordinates,
   draw the graph of $y = x - 2$.

2. Write down the gradient and intercept of
   each of these straight-line graphs.
   a) $y = 4x - 1$ **b)** $y = 3 - 2x$ **c)** $2y = 4x + 8$

3. Match each of the three graphs below with
   one of the following equations:
   * $y = 2x - 5$
   * $y = x^2 + 3$
   * $y = 3 - x^2$
   * $y = 5 - x$
   * $y = x^3$

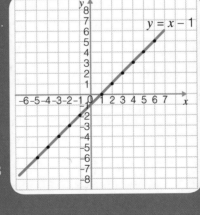

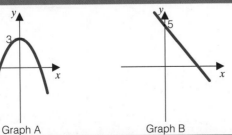

Graph A   Graph B   Graph C

**Progress
Check**

11 **4.3** Interpreting graphical information

It is important that you can interpret graphical information from a variety of situations.

Using linear graphs

Make a collection of graphs from newspapers and magazines. Practise interpreting them. Look at the scales used and the sorts of information being compared.

Linear graphs are often used to show relationships. For example:

Neville has a window-cleaning round. He charges £30 for the use of his materials and £15 per hour after that. This information can be put into a table.

No. of hrs	0	1	2	3	4
Charge (£)	30	45	60	75	90

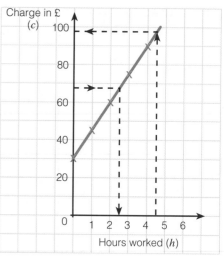

The equation of this graph is $C = 15h + 30$. The gradient is 15 (i.e. the charge per hour) and the intercept is (0, 30), i.e. the standing charge for his materials.

The graph of this information shows that there is a linear relationship.

Use a national newspaper or the Internet to find out the different exchange rates between Pounds Sterling and other currencies. Using graph paper, make your own conversion graphs to show the exchange rates. Use your graphs to convert different amounts of a particular currency.

The graph can be used to find, for example, how long Neville works if he charges £67.50 (2.5 hours) and what he charges if he works 4.5 hours (£97.50).

Conversion graphs

Conversion graphs are used to change one unit of measurement into another unit; for example, litres to pints, kilometres to miles, pounds to dollars, etc.

For example:

£1 = $1.50

Measurements such as kilometres and miles are in direct proportion to each other. For example, 8km = 5 miles. A conversion graph of this information can be drawn. See Topic 7.1.

To change dollars to pounds, read across to the line and then read down: for example, $4 is £2.67 (approx.)

To change pounds to dollars, read up to the line then read across; for example, £4.50 is $6.80 (approx.)

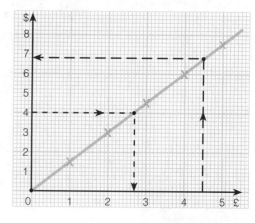

Distance–time graphs

Distance–time graphs are often known as travel graphs. Distance is on the vertical axis; time is on the horizontal axis. The speed of an object can be found on a distance–time graph by using:

$$\text{Speed} = \frac{\text{distance}}{\text{time}}$$

Speed is a compound measure because it involves a combination of two basic measures. See Topic 7.1.

Example

The travel graph shows the car journeys of two people.

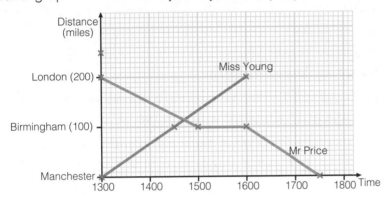

From the travel graph find the following:

a) The speed at which Miss Young is travelling.

$$\text{Speed} = \frac{\text{distance}}{\text{time}} = \frac{200}{3} = 66.7\text{mph (1 d.p.)}$$

b) The length of time Mr Price had a break.
Mr Price is stationary between 1500 and 1600, i.e. 1 hour.

c) The speed of Mr Price from Birmingham to Manchester.

$$\text{Speed} = \frac{\text{distance}}{\text{time}} = \frac{100}{1.5} = 66.7\text{mph (1 d.p.)}$$

d) The time at which Miss Young and Mr Price pass each other.
Since each small square is 6 minutes, they pass at 14.42.

e) The speed of Mr Price from London to Birmingham.

$$\text{Speed} = \frac{\text{distance}}{\text{times}} = \frac{100}{2} = 50\text{mph}$$

The above example highlights the following key points.

The steeper the graph, the greater the speed. Object A is travelling faster than object B, which in turn is travelling faster than object C.

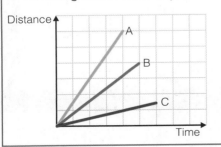

The green line shows an incorrect journey time because you cannot go back in time.

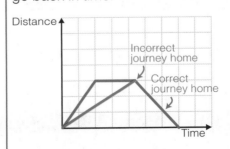

Make graphs from your own journeys. Play around with a variety of scales so that you become accustomed to which scale would be appropriate to a particular set of data.

Matching graphs to real-life situations

Example

These containers are being filled with water at a rate of 150ml per second. The graphs show how the depth of the water changes with time. Match the containers with the correct graphs.

Container A is graph 3 since the depth of the water changes uniformly with time.

Container B is graph 1 since the depth will rise quickly in the narrow part of the cone and then begin to slow down.

Container C is graph 2 since the depth will increase slowly at the wider part of the container and then increase more quickly at the narrow part.

Progress Check

1. The distance–time graph shows Mrs Robert's car journey.
   a) At what speed did she travel for the first 2 hours?
   b) What is happening at A?
   c) At what speed is her return journey?

2. The graph shows the charges made by a van-hire firm.
   a) What do you think point A represents?
   b) By using the gradient of the line, work out how much was charged per day for the hire of the van.
   c) Write down a formula that connects the cost (C) of the van hire and the number of days (d).

3. Water is poured into these containers at a constant rate. Match each container to the correct graph.

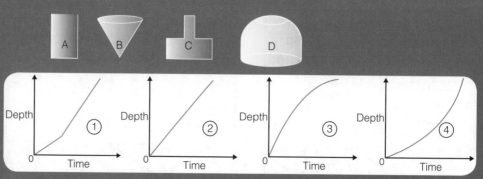

Worked questions

First check that the difference between the numbers is constant. The formula for finding the nth term is $T(n) = a(n) +/- b$ where a = the difference between the terms and b = the difference between a and the first term.
So, $a = 5$ and $b = -3$

In the equation in question 2, $a = 3$ and $b = +2$

To find the next terms, add 3.

Always check your answer using the equation:

$T(4) = (3 \times 4) + 2 = 14$

1. Find the nth term in this sequence: 2, 7, 12 *(1 mark)*

The nth term is $5n - 3$

2. The nth term of a sequence is $3n + 2$.

a) Find the first four terms of the sequence. *(1 mark)*

Term 1 is $(3 \times 1) + 2 = 5$

Term 2 = $5 + 3 = 8$
Term 3 = $8 + 3 = 11$
Term 4 = $11 + 3 = 14$

The first four terms are 5, 8, 11, 14

b) Find the 50th term. *(1 mark)*

$T(50) = (3 \times 50) + 2 = 152$

3. Stars are used to make patterns:

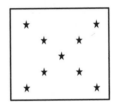

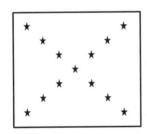
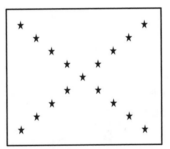

a) How many stars are required for the nth pattern? *(1 mark)*

Questions often show sequences in a practical way but the same rules apply. Check that the differences are the same. It is a good idea to make a chart.

Term 1	Term 2	Term 3	Term 4	T(n)
5	9	13	17	4(n) + 1

4	4	4

The difference between the terms is constant and equals 4 so, $a = 4$ and $b = +1$

The number of stars required for the nth pattern is $4n + 1$.

b) How many stars will be needed to make the 12th pattern? *(1 mark)*

The number of stars needed for the 12th pattern is $(4 \times 12) + 1 = 49$

First of all, draw up a set of coordinates by substituting different values for x. You are always advised to choose a minimum of three values for x.

Now plot your points on the graph and join them with a straight line, ensuring that you draw clearly through both axes.

4. Draw a set of axes from –10 to 14 on the y-axis and from –4 to 4 on the x-axis.

a) Draw the graph of $y = 4x + 2$ using the substitution method.

(2 marks)

When $x = 0$, $y = 2 \rightarrow (0, 2)$

When $x = 3$, $y = 14 \rightarrow (3, 14)$

When $x = -3$, $y = -10 \rightarrow (-3, -10)$

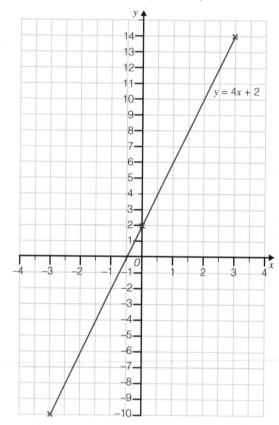

Before you attempt this method, ensure that the equation is in the form of $y = mx + c$.

–1 is the intercept value on the y-axis.

In other words, this is the value of y at the point where the line crosses the y-axis. Put a dot on the y-axis at this value.

Now look at the gradient value – the coefficient of x. From the intercept point, go right 1 unit and up 4 units. (If the coefficient is negative, i.e. $-4x$, you would go down four units.) Put a dot at this value. Repeat this at least twice more, then join up the points as before.

b) Draw a set of axes from –7 to 11 on the y axis and from –4 to 4 on the x axis. Draw the graph of $y = 4x - 1$ using the gradient/intercept method.

(2 marks)

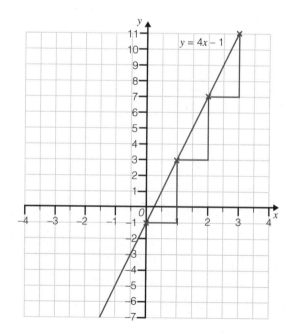

c) Find the equation of the line shown here: *(1 mark)*

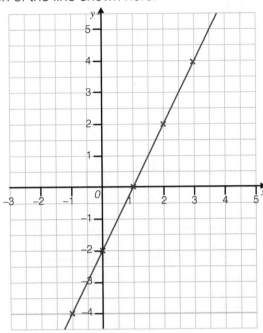

Remember that equations for straight lines should be given in the form of $y = mx + c$.

To measure the gradient, divide the vertical distance between two points by the horizontal distance between them. (If the line slopes upwards from right to left the gradient will be negative.) Pick the point (1, 0). To get to the next point (2, 2) you would go right one unit and up two units (2 ÷ 1 = 2).

The intercept is −2 so $c = -2$
The gradient is 2 so $m = 2$.
The equation for this straight line is $y = 2x - 2$.

5. This graph shows the cost of Graham's water usage.

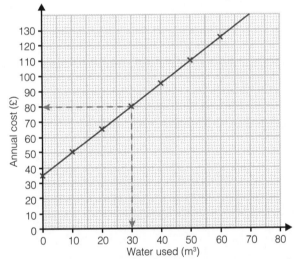

a) If Graham uses 30m³ of water, what will be the cost? *(1 mark)*

The cost for 30m³ is £80.

a) Draw ruled dotted lines on the graph to work this out.

b) What is the fixed or standing charge? *(1 mark)*

The fixed charge is £35.

b) This is shown by the y-intercept.

c) What is the gradient of the line? *(1 mark)*

The horizontal difference is 10 and the vertical difference is 15.

The gradient is $1\frac{1}{2}$.

c) Take care when finding the gradient on graphs like this because the scales are likely to be different.

$15 \div 10 = 1\frac{1}{2}$

d) Write down the equation for the cost of a year's water in terms of the amount used. *(1 mark)*

Cost = £$1\frac{1}{2}x$ + 35.

d) Write in the format $y = mx + c$

e) Use this equation to work out the cost of 84m³ of water. *(1 mark)*

Cost 5 £$(1\frac{1}{2} \times 84)$ + 35 = £126 + 35 = £161

The cost for 84m³ is £161.

93

Practice questions

1.   Here are four sets of coordinates:

   (–2, 3)   (4, –5)   (–1, –1)   (4, 3)

   a) Which pair, when joined, would make a horizontal line?   *(1 mark)*

   b) Which pair, when joined, would make a vertical line?   *(1 mark)*

   c) Which pair, when joined, would make a positive gradient?   *(1 mark)*

   d) Which pair, when joined, would make a negative gradient?   *(1 mark)*

2.
   P: $y - 3 = 2x$ Q: $2y = 4x + 6$ R: $2y = 3x$

   S: $2x + 2y = 7$ T: $y = 3 - 2x$

   a) Which of these lines are parallel?   *(2 marks)*

   b) Which of these lines goes through (0, 0)?   *(1 mark)*

   c) Which of these lines have a negative gradient?   *(2 marks)*

   d) Which of these lines go through the point (2, 7)?   *(2 marks)*

3.   This question is about converting Pounds Sterling to American dollars. ▣

   a) On the squared paper provided, draw a conversion graph for converting
   Pounds Sterling to American dollars. Mark the vertical axis £ and the
   horizontal axis $. The scale for £s should increase in 10s and the
   scale for $s in 20s. On a certain day, the rate was £1 = $1.6.   *(3 marks)*

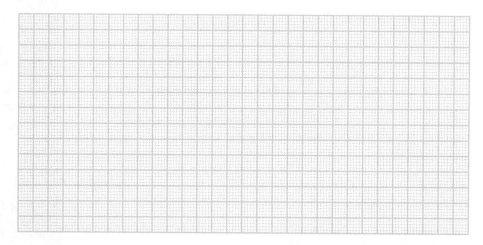

   b) By drawing lines on your graph, show how many dollars you would get
   for £55, £37.50 and £15.   *(3 marks)*

4. Maria took her dog, River, for a walk. Her journey is shown on the graph below.
The letters show the various stages of her walk.

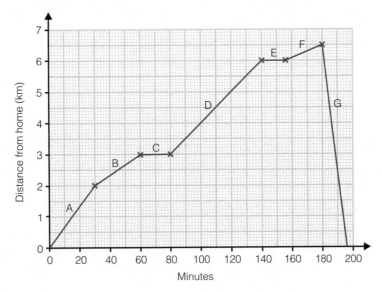

a) Which letters represent the hills on Maria's walk? *(2 marks)*

b) Maria stopped to eat her lunch for 20 minutes. Which letter represents this time? *(1 mark)*

c) Maria also stopped to chat. How long did her conversation last? *(1 mark)*

d) What happened at G? How could the steepness of the line be explained? *(2 marks)*

5. Gordon packs chocolate bars for a living. He earns an extra £10 for every
15 boxes he packs. This graph illustrates what Gordon's weekly wage could be.

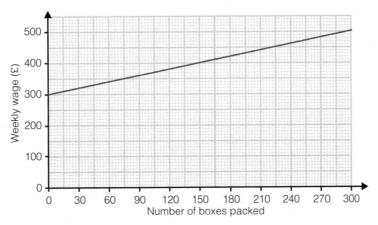

a) What is Gordon's basic wage? *(1 mark)*

b) Last week Gordon packed 210 boxes. What was his wage? *(1 mark)*

c) What is the gradient of the line? *(1 mark)*

d) Write down the equation for Gordon's wage in terms of the number of boxes
he packs. *(1 mark)*

e) Use your equation to work out what Gordon's wage would be if he packed
420 boxes. *(1 mark)*

f) Next week, Gordon wants to earn £720. Use your equation to work out how
many boxes he needs to pack. *(1 mark)*

6. Complete the table.

Sequence					n^{th} term
6	11	16	21	26	$5n + 1$
8	17	26	35	44	
5					$3n + 2$
2	8	18	32	50	
1					$n^2 + 2n - 2$

(4 marks)

7. Look at the pattern of rings below.

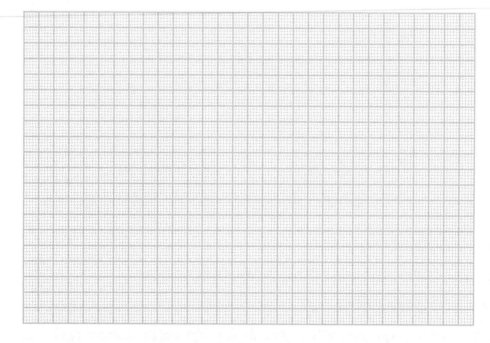

a) How many rings are required for the nth pattern? (1 mark)

b) How many rings will be needed to make the 40th pattern? (1 mark)

8. **a)** Draw and complete a table of values for $y = x^2 + 2x - 2$ for values of x from –5 to 3. (4 marks)

b) On the grid provided, draw a set of axes from –4 to 14 on the y-axis and from –6 to 6 on the x-axis. Draw the graph of $y = x^2 + 2x - 2$. (2 marks)

c) By using the graph, solve the equation $x^2 + 2x - 2 = 0$

Give your answer to the nearest tenth. (2 marks)

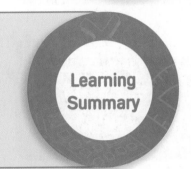

After studying this chapter you should be able to:
- use a wide range of properties of two- and three-dimensional shapes
- classify quadrilaterals by their geometric properties
- understand and use congruence
- use angle properties, bearings and scale drawings
- use and apply Pythagoras' theorem when solving problems in 2-D
- use and apply trigonometry in right-angled triangles when solving problems.

Learning Summary

5.1 Two- and three-dimensional shapes

Lines

A straight line is one-dimensional. It only has length.

A **line segment** is of finite length. For example, the line segment PQ has endpoints P and Q.

P   Q

You need to know these properties of lines.

> Two lines are **parallel** if they are in the same direction. They are always equidistant (the same distance apart).

> Two lines are **perpendicular** if they are at right angles to each other. Perpendicular lines meet at 90°.

Two-dimensional shapes

Two-dimensional (2-D) shapes have area. All points on 2-D shapes are in the same plane.

Below are the 2-D shapes you need to recognise along with some of their important properties. When two lengths have a line with a dash on them, it means they are equal in length.

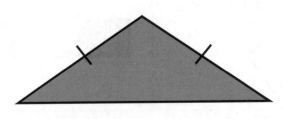

Triangles

Triangles have three sides. There are several types of triangle:

Right-angled	Equilateral
Has a 90° angle.	Three sides equal. Three angles equal.
Isosceles	**Scalene**
Two sides equal. Base angles equal.	All the sides and angles are different. 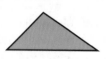

Quadrilaterals

Quadrilaterals have four sides. There are several types of quadrilateral:

Square
- Four lines of symmetry
- Rotational symmetry of order 4
- All angles are 90°
- All sides equal
- Two pairs of parallel sides
- The diagonals are equal and **bisect** each other at right angles.

Rectangle
- Two lines of symmetry
- Rotational symmetry of order 2
- All angles are 90°
- Opposite sides equal
- Two pairs of parallel sides
- The diagonals bisect each other

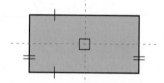

Parallelogram
- No lines of symmetry
- Rotational symmetry of order 2
- Opposite sides are equal and parallel
- Opposite angles are equal

Rhombus
- Two lines of symmetry
- Rotational symmetry of order 2
- All sides are equal
- Opposite sides are parallel
- Opposite angles are equal
- The diagonals bisect each other at right angles and also bisect the corner angles

Kite
- One line of symmetry
- No rotational symmetry
- Diagonals do not bisect each other
- Two pairs of adjacent sides are equal
- Diagonals cross at right angles

Trapezium
- Has one pair of parallel sides
- No lines of symmetry
- No rotational symmetry
- An isosceles trapezium has one line of symmetry

Properties of different triangles are used when finding missing angles. See Topic 5.2.

Create these different types of triangle using straws to represent the sides. Cut the straws to create any shorter sides.

Copy each of these quadrilateral properties onto separate cards. Make a sketch of each quadrilateral and then match the properties to the sketch. Ultimately, you should be able to recite them without any help.

Polygons

Polygons are 2-D shapes with straight sides. Regular polygons are shapes with all sides and angles equal.

Number of sides	Name of polygon
3	Triangle
4	Quadrilateral
5	Pentagon
6	Hexagon
7	Heptagon
8	Octagon
9	Nonagon
10	Decagon

> The properties of regular polygons mean that all the interior angles and all the exterior angles are the same. See Topic 5.2.

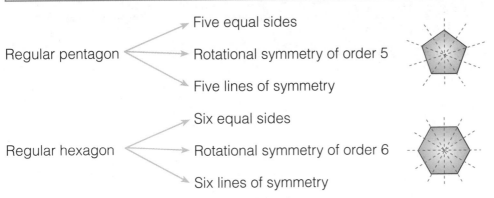

Regular pentagon → Five equal sides
→ Rotational symmetry of order 5
→ Five lines of symmetry

Regular hexagon → Six equal sides
→ Rotational symmetry of order 6
→ Six lines of symmetry

> Make rough sketches of the different types of polygon. Draw the lines of symmetry on them. On separate sketches, draw lines from the centre of the polygon to each point. Then, again on separate sketches join the points to see how many triangles you can make. It will always be 2 less than the number of sides.

The circle

You need to know these properties of circles.

Circumference – the distance around the outside edge of a circle **Diameter** – the distance of a straight line passing through the centre of the circle from one side to the other **Radius** – half the length of the diameter	Circumference, Radius, O, Diameter
Chord – a line that joins two points on the circumference without going through the centre of the circle **Arc** – part of the circumference	Chord, Tangent, Arc, O
Tangent – a line that touches the circle at one point only The radius and tangent make an angle of 90° at the point they meet.	O, Radius, Tangent
The **perpendicular bisector** of a chord passes through the centre of a circle.	O, Perpendicular bisector
The angle in a semicircle is always a right angle.	O

> The perpendicular bisector can be accurately constructed with compasses. See Topic 6.2.

Three-dimensional solids

Three-dimensional (3-D) objects have volume (or **capacity**). You should know these 3-D solids:

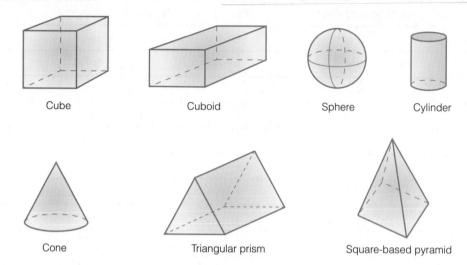

Cube   Cuboid   Sphere   Cylinder

Cone   Triangular prism   Square-based pyramid

A **prism** is a solid which can be cut into slices that are all the same shape.

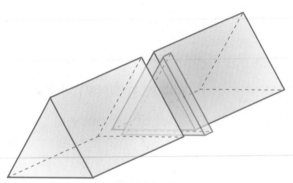

You show know these features of 3-D solids:

A **face** is a flat surface of a solid.	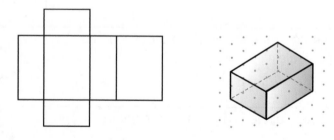
An **edge** is where two faces meet.	
Vertex is another word for corner.	A cuboid has 6 faces, 8 vertices and 12 edges.

> Take a shoe box (or other small box) and identify the faces, edges and vertices on it.

The **net** of a 3-D solid is a 2-D (flat) shape that can be folded to make the 3-D solid. The net of the cuboid would look like this:

You can represent 3-D shapes on isometric paper. On this paper you can draw lengths in three perpendicular directions on the same scale. The faces do not appear in their true shape. A 'T'-shaped prism can be shown clearly on isometric paper.

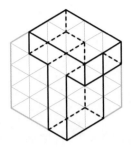

Plans and elevations

A **plan** is what can be seen if a 3-D solid is looked down on from above. Architects often use plans to show the design of new properties. An **elevation** is seen if the 3-D solid is looked at from the side or front.

Try sketching the front and side elevations of a variety of buildings near to your house. Make some models with building blocks and sketch all the elevations.

Example

Draw a sketch of the plan and the elevations from A and B of this solid.

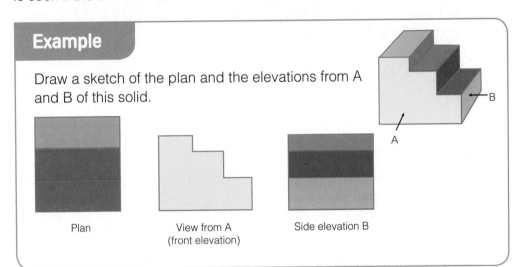

| Plan | View from A (front elevation) | Side elevation B |

Coordinates in three dimensions

The normal x-y coordinates can be extended into a third direction, known as z. All positions have three coordinates (x, y, z). Remember to read them in the order (x, y, z).

Coordinates can be either positive or negative numbers. See Topic 1.1.

In this cuboid, the vertices have the following (x, y, z) coordinates:

A (3, 0, 0)

B (3, 2, 0)

C (0, 2, 0)

D (0, 2, 1)

E (0, 0, 1)

F (3, 0, 1)

G (3, 2, 1)

O (0, 0, 0)

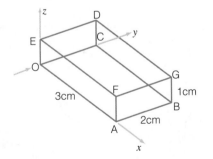

Symmetry

There are three different types of symmetry.

Reflective symmetry

Reflective symmetry is when both sides of a shape are the same on each side of a mirror line. The mirror line is known as a line or axis of symmetry. For example:

| 1 line of symmetry | 1 line of symmetry | 3 lines of symmetry | No line of symmetry |

Example

Half a reflected shape is shown here. The dashed line is the line of symmetry. Copy and complete this shape.

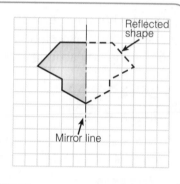

Rotational symmetry

A 2-D shape has rotational symmetry if, when it is turned, it looks exactly the same. The order of rotational symmetry is the numbers of times the shape can be turned and still look the same.

For example:

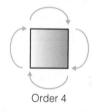

Order 4   Order 2   Order 1 (no rotational symmetry)

For the kite, there is one position. It is said to have rotational symmetry of order 1 or no rotational symmetry.

Plane symmetry

Plane symmetry only exists in 3-D solids. A 3-D solid has a plane of symmetry if the plane divides the shape into two halves, and one half is the exact mirror image of the other. 3-D solids can have more than one plane of symmetry.

For example:

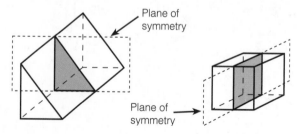

When drawing in a plane of symmetry, you must show its edges.

Draw all the capital letters and numbers from 1 to 9. Place a small handbag mirror on its edge at the side of each character. Take a good look at the mirror image and then draw it at the side. Then try to work out the order of rotation of the letters and numbers.

Congruent shapes

Shapes are **congruent** if they are exactly the same size and shape, i.e. they are identical. Two shapes are congruent even if they are mirror images of each other.

Triangles are congruent if one of the following sets of conditions is true:
(S stands for side, A for angle, R for right angle, H for hypotenuse.)

Condition	Description	Example
SSS	The three sides of one triangle are the same as the three sides of the other triangle.	
SAS	Two sides and the angle between them in one triangle are equal to two sides and the angle between them in the other triangle.	
RHS	Each triangle contains a right angle. Both hypotenuses and another pair of sides are equal.	
AAS	Two angles and a side in one triangle are equal to two angles and the corresponding side in the other.	

The hypotenuse is used in Pythagoras' theorem. See Topic 5.3.

Make individual revision cards with these descriptions of congruency on them. Practise reciting them off by heart.

1. Draw an accurate net of this 3-D shape.

   5.7 cm
   4 cm
   4 cm
   4 cm

2. Draw a plane of symmetry on this solid.

3. Are these two triangles congruent? Explain why.

   5cm 25° 7cm
   7cm 25° 5cm

4. Hywel shades in a shape made of five squares on a grid.
   a) Shade in one more square to make a shape with line A as its line of symmetry. Call the square R.
   b) Going back to the original shape, shade in two more squares to make a shape that has line B as a line of symmetry. Call the squares S and T.

   Line B
   Line A

5. For the solid shown, draw:
   a) the plan
   b) the view from A
   c) the view from B.

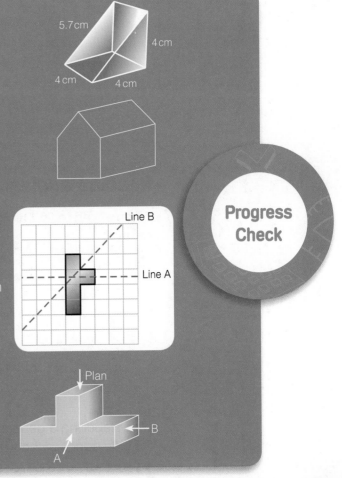

   Plan
   B
   A

Progress Check

5.2 # Angles, bearings and scale drawings

Angles and the protractor

An **angle** is the amount of turning or rotation. Angles are measured in **degrees**. A circle is divided into 360 parts. Each part is called a degree and is represented by a small circle, °.

Acute	Obtuse	Reflex	Right angle
An angle between 0° and 90°.	An angle between 90° and 180°.	An angle between 180° and 360°.	An angle of 90°.

A protractor is used to measure the size of an angle:

Make sure you put 0° at the start position, and that you read from the correct scale.

For this angle, measure on the outer scale since you must start at 0°.

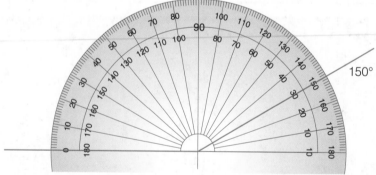

150°

This angle is 150°.

Reading angles

When asked to find angle $A\hat{B}C$ or $\angle A\hat{B}C$, find the middle letter angle, i.e. at B:

$\angle ABC = 30°$

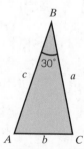

When labelling a general triangle the side opposite the vertex A is called a, the side opposite the vertex B is called b and the side opposite the vertex C is called c.

Angle facts

There are some facts about angles that you need to know:

Angles on a straight line add up to 180°.	$a + b + c = 180°$
Angles in a triangle add up to 180°.	$a + b + c = 180°$
Vertically opposite angles are equal.	
Angles at a point add up to 360°.	$a + b + c = 360°$
Angles in a quadrilateral add up to 360°.	$a + b + c + d = 360°$
An exterior angle of a triangle equals the sum of the two opposite interior angles.	$c = a + b$ Since if: $a + b + d = 180°$ (angles in a triangle add up to 180°) $d + c = 180°$ (angles on a straight line add up to 180°) Then $c = a + b$

Example

Find the angles labelled by letters.

a)

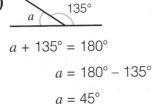

$a + 135° = 180°$

$a = 180° - 135°$

$a = 45°$

b)

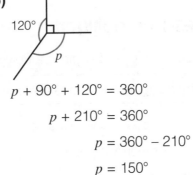

$p + 90° + 120° = 360°$

$p + 210° = 360°$

$p = 360° - 210°$

$p = 150°$

c)

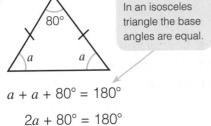

In an isosceles triangle the base angles are equal.

$a + a + 80° = 180°$

$2a + 80° = 180°$

$2a = 180° - 80°$

$2a = 100°$

$a = 50°$

d)

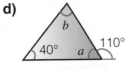

$a + 110° = 180°$

$a = 70°$

$40° + b = 110°$

$b = 110° - 40°$

$b = 70°$

Algebra is used to set up an equation in order to calculate the missing angle. See Topic 3.2.

Angles in parallel lines

You need to know these rules about angles in parallel lines:

Alternate angles are equal.	
Corresponding angles are equal.	
Allied or supplementary angles add up to 180°.	$c + d = 180°$

For example:

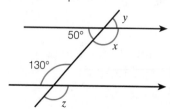

$x = 130°$ (alternate)

$y = 50°$ (vertically opposite) angles are equal

$z = 130°$ (vertically opposite) angles are equal

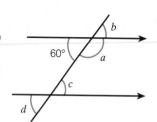

$a = 120°$ (angles on a straight line) add to 180°

$b = 60°$ (vertically opposite) angles are equal

$c = 60°$ (alternate) angles are equal

$d = 60°$ (vertically opposite to c)

> Draw your own sets of parallel lines and label angles which are equal and those which make 180°.

Angles in a polygon

There are two types of angle in a polygon: interior (inside) and exterior (outside).

> For any polygon with n sides → sum of exterior angles = 360°

For a regular polygon with n sides:

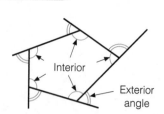

- size of exterior angle = $\dfrac{360°}{n}$
- interior angle + exterior angle = 180°
- sum of interior angles = $(n - 2) \times 180°$

Any polygon can be split up into triangles. If a polygon has n sides, then $(n - 2)$ triangles can be made. Since each triangle adds up to 180°, the sum of the interior angles = $(n - 2) \times 180°$

Example

1. Calculate the interior and exterior angle of a regular hexagon. A hexagon has six sides.

   Exterior angle $= \dfrac{360°}{6} = 60°$

   Interior angle $= 180° - 60°$

   $\qquad\qquad\;\; = 120°$

2. Find the sum of the interior angles of a regular pentagon.

   A pentagon has five sides.

   Sum of interior angles $= (n - 2) \times 180°$

   $\qquad\qquad\qquad\qquad\; = (5 - 2) \times 180°$

   $\qquad\qquad\qquad\qquad\; = 3 \times 180°$

   $\qquad\qquad\qquad\qquad\; = 540°$

It is important that you know the names of different polygons and how many sides they have. See Topic 5.1.

Tessellations

A **tessellation** is a pattern of 2-D shapes that fit together without leaving any gaps. Tessellations are seen throughout the history of art, from ancient to modern art. They frequently appear in the art of M.C. Escher.

Here are some examples of tessellations:

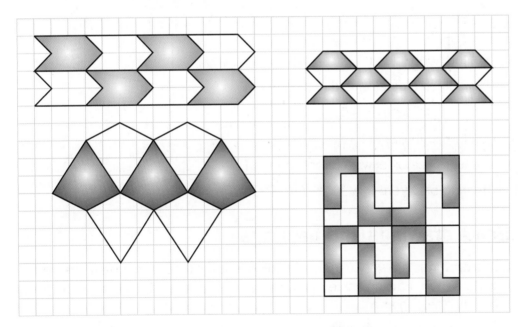

To work out whether shapes tessellate, you need to be able to calculate the size of the interior angle of the shape. See Topic 5.2.

For shapes to tessellate, the angles at each point must add up to 360°.

Compass directions and bearings

The diagram shows the points of the compass. Directions can also be given as **bearings**. Bearings are used on aeroplanes and ships to make sure they are travelling in the right direction and to avoid collisions.

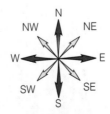

Bearings give directions in degrees. They are always measured from the north in a clockwise direction. A bearing must have three figures. The word 'from' indicates the position of the north line from which the angle is measured.

For example:

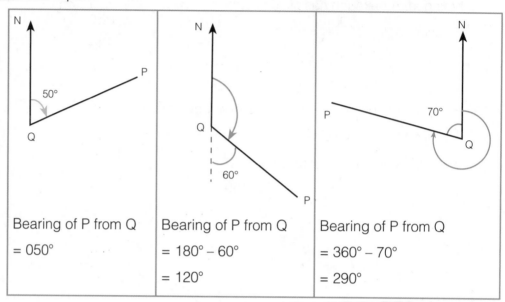

Bearing of P from Q	Bearing of P from Q	Bearing of P from Q
= 050°	= 180° − 60°	= 360° − 70°
	= 120°	= 290°

Since we are finding the bearing of P from Q, the north line is placed at Q. The bearing is measured in a clockwise direction from this north line.

When finding a back bearing (the bearing of Q from P in the diagrams above):
* Draw a north line at P.
* The two north lines are parallel lines, so the angle properties of parallel lines are used.

For example:

* Put the north line at P.

  Measure in a clockwise direction from P.

  Bearing of Q from P is 50° + 180° = 230°

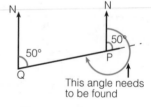

* Put the north line at P.

  Measure in a clockwise direction from P.

  Bearing of Q from P is 360° − 60° = 300°

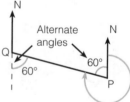

Scale drawings and bearings

Scale drawings are very useful for finding lengths that cannot be measured directly. When drawing scale diagrams, the lengths need to be accurate to 2mm and the angles to 2°.

Example

A ship sails from a harbour for 15km on a bearing of 040°, and then continues due east for 20km. Make a scale drawing of this journey using a scale of 1cm to 5km. How far will the ship have to sail to get back to the harbour by the shortest route? What will the bearing be?

Shortest route = 6.4km × 5km   Bearing = 70° + 180°

   = 32km   = 250°

N
N

Ship

Bearing
= 180° + 70°
= 250°

20km

15km

70°   180°

40°

Shortest route = 6.4 × 5km = 32km

Harbour

This diagram is not drawn to scale but is used to show what your diagram should look like.

Maps and diagrams

Scales are often used on maps and diagrams. They are usually written as ratios.

Examples

1. The scale on a road map is 1 : 25 000. Manchester and Rochdale are 60cm apart on the map. Work out the real distance between them in km.

   On a scale of 1 : 25 000, 1cm on the map represents 25 000cm on the ground.

   60cm represents 60 × 25 000 = 1 500 000cm.

   1 500 000 ÷ 100 = 15 000m   ← Divide by 100 to change cm to m.

   15 000 ÷ 1000 = 15km   ← Divide by 1000 to change m to km.

   The distance between Manchester and Rochdale is 15km.

2. A house plan has a scale of 1 : 30. If the width of the house on the plan is 64cm, what width is the real house?

   1cm represents 30cm.

   64cm represents 64 × 30 = 1920cm

   1920 ÷ 100 = 19.2m

   The width of the house is 19.2m.

Scales are often written as ratios, so you need to be able to calculate with ratios when quantities are in proportion. See Topic 1.4.

Looking at Ordnance Survey maps is a good way to get to grips with scale. Also study road maps. Use the scale to work out the real distances between two locations. You could also draw scale plans of your own bedroom or your garden.

Progress Check

1. All the angles in an equilateral triangle are 60°. True or false?

2. Calculate the size of the angles labelled with letters.

a)

b)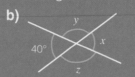

c)

d)

3. The scale on a road map is 1 : 50 000. If two towns are 14cm apart on the map, work out the real distance in km between them. 📖

4. The diagram shows three legs of a cross-country course. The course starts at T, then goes to R and then P, and finally back to T.

   a) Find the bearing of R from T.

   b) Find the bearing of R from P.

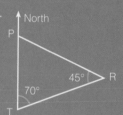

5. Calculate the size of the interior angle of a regular pentagon. 📖

 5.3 # Pythagoras' theorem

Pythagoras' theorem is used to calculate the length of the third side of a right-angled triangle, when the other two sides are known. You need to know the formula for Pythagoras' theorem and how it can be rearranged to give the formulae for calculating the lengths of the shorter sides.

The **hypotenuse** is the longest side of a right-angled triangle. It is always opposite the right angle.

Pythagoras' theorem states that:

> In any right-angled triangle, the square on the hypotenuse is equal to the sum of the squares on the other two sides.

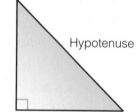

Using the letters in the diagram below, the theorem is written as:

$c^2 = a^2 + b^2$

This can be rearranged to give:

$b^2 = c^2 - a^2$

or

$a^2 = c^2 - b^2$

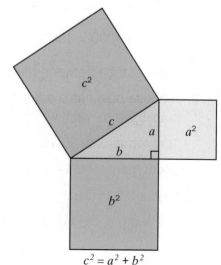

$c^2 = a^2 + b^2$

Example

Find the length of the missing side in each of these triangles, giving your answer to 1 decimal place. 📟

a)

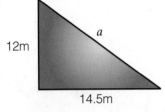

$a^2 = 12^2 + 14.5^2$

$a^2 = 354.25$

$a = \sqrt{354.25}$ ← If you are not told to what degree of accuracy to round, be guided by significant figures in the question.

$a = 18.8m$ (1 d.p.)

b)

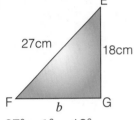

$27^2 = b^2 + 18^2$

$b^2 = 27^2 - 18^2$

$b^2 = 405$

$b = \sqrt{405}$

$b = 20.1cm$ (1 d.p.)

c) Prove that this triangle is right-angled.

By Pythagoras' theorem:

$10^2 = 6^2 + 8^2$

$100 = 36 + 64$

$100 = 100$

Since this is correct and obeys Pythagoras' theorem, then the triangle must be right-angled.

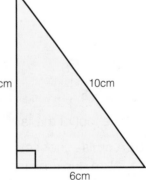

Solving using Pythagoras' theorem is simply substituting into a formula. See Topic 3.1.

Calculating the length of a line segment

You can calculate the length of the line segment joining two points by using Pythagoras' theorem.

For example, by drawing in a triangle between the two points A(1, 2) and B(7, 6), the length of AB can be found by Pythagoras' theorem.

Horizontal distance = 7 – 1 = 6

Vertical distance = 6 – 2 = 4

Length of $(AB)^2 = 6^2 + 4^2$

$(AB)^2 = 36 + 16$

$(AB)^2 = 52$

AB = $\sqrt{52}$ ← We could leave the answer in surd form.

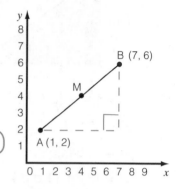

> Answers can be left in surd form, if a calculator is not allowed. See Topic 1.1.

Length of AB = 7.21 (2 d.p.)

The midpoint of AB, M, can also be found. M has coordinates of (4, 4), i.e.

$$\left(\frac{1+7}{2}, \frac{2+6}{2}\right)$$

Solving problems

Pythagoras' theorem can be used to solve practical problems.

Example

a) A ladder of length 13m rests against a wall. The ladder reaches 12m up the wall. How far away from the wall is the foot of the ladder?

$13^2 = x^2 + 12^2$

$x^2 = 13^2 - 12^2$

$x^2 = 25$

$x = \sqrt{25}$

$x = 5$

The foot of the ladder is 5m away from the wall.

b) A cruise liner sets sail from Port A and travels 80km due east then 50km due north, to reach Port B. How far is Port A from Port B by the shortest route? 🖩

$a^2 = 80^2 + 50^2$

$a^2 = 6400 + 2500$

$a^2 = 8900$

$a = \sqrt{8900}$

Shortest route = 94.3 km (3 s.f.)

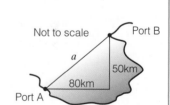

Not to scale

Port B

Port A

a

50km

80km

1. Calculate the lengths of the sides marked with a letter.

   Give your answers to 3 s.f.

   a)

   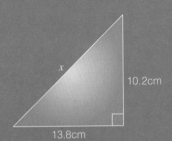

   b)

   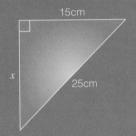

Progress Check

2. Calculate the height of this isosceles triangle.

   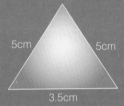

3. a) The coordinates of two points are (1, 2) and (7, 10).

   What is the length of the line joining these two points?

   b) Find the coordinates of the midpoint of the line joining the two points.

5.4 # Trigonometry in right-angled triangles

Trigonometric ratios

In a right-angled triangle the sides and the angles are related by three trigonometrical ratios: the sine (abbreviated to sin), the cosine (abbreviated to cos) and the tangent (abbreviated to tan).

To use these ratios, you first need to be able to label the sides of the triangle:
* hyp (hypotenuse) is opposite the right angle.
* opp (opposite side) is opposite the angle θ
* adj (adjacent side) is next to the angle θ

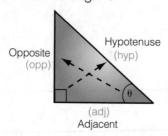

θ is a Greek letter called theta and is used to represent angles.

$$\sin θ = \frac{\text{opposite}}{\text{hypotenuse}}$$

$$\cos θ = \frac{\text{adjacent}}{\text{hypotenuse}}$$

$$\tan θ = \frac{\text{opposite}}{\text{adjacent}}$$

> Make cards with each of the letters SOHCAHTOA on them. Mix them up and practise arranging them back into the correct order.

The made-up word SOH CAH TOA is a quick way of remembering the ratios.

sin equals opposite divided by hypotenuse	cos equals adjacent divided by hypotenuse	tan equals opposite divided by adjacent

Trigonometry is used to calculate the length of a missing side and the size of a missing angle in right-angled triangles.

Examples

1. Calculate the length of *BC*.

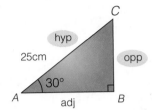

Label the sides first.

Decide on the ratio:

$$\sin 30° = \frac{\text{opp}}{\text{hyp}}$$

Substitute in the values:

$$\sin 30° = \frac{BC}{25}$$

$$25 \times \sin 30° = BC$$

$$BC = 12.5\text{cm}$$

> You are required to substitute into the correct formula. See Topic 3.1.

2. Calculate the length of *EF*.

$$\cos 40° = \frac{\text{adj}}{\text{hyp}}$$

$$\cos 40° = \frac{20}{EF}$$

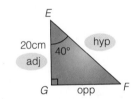

$EF \times \cos 40° = 20$ ←   Multiply both sides by *EF*.

$EF = \dfrac{20}{\cos 40°}$ ←   Divide both sides by cos 40°.

$$EF = 26.1\text{cm (1 d.p.)}$$

> It is important that you are familiar with how to use your calculator for trigonometry problems. See Topic 2.1.

Using a calculator key in:

    or

3. Calculate angle ABC.

Label the sides and decide on the ratio.

$$\tan \theta = \frac{\text{opposite}}{\text{adjacent}}$$

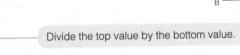

$\tan \theta = \dfrac{15}{27}$ ←   Divide the top value by the bottom value.

$\tan \theta = 0.\dot{5}$

$\theta = \tan^{-1} 0.\dot{5}$ ←   The tan⁻¹ shows that we need the angle whose tangent is 0.5̇

$\theta = 29°$ ←   To the nearest degree.

Solving problems using trigonometry

Trigonometry can be used to solve problems involving right-angled triangles.

Angle of elevation	Angle of depression
The angle of elevation is measured from the horizontal upwards.	The angle of depression is measured from the horizontal downwards.
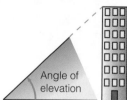 Angle of elevation	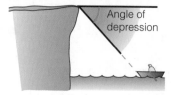 Angle of depression

Sketch your own right-angled triangles. Put in two values (e.g. a side and an angle, or two sides) and work out how you would calculate a third value. Keep practising with your calculator so that using sin, cos and tan become second nature.

Examples

1. Dipak stands 30m from the base of a tower. He measures the angle of elevation from ground level to the top of the tower as 50°. Calculate the height of the tower. Give your answer to 3 s.f.

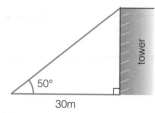

$$\tan 50° = \frac{\text{opp}}{\text{adj}} = \frac{\text{height}}{30}$$

$30 \times \tan 50° = \text{height of tower}$

Height of tower = 35.8m (3 s.f.)

> Make sure you give your answers to the correct degree of accuracy.

2. Siân is flying a kite. The string is 30 m long and is at an angle of 40° to the horizontal. How high is the kite above Siân's head?

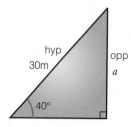

$$\sin 40° = \frac{\text{opp}}{\text{hyp}}$$

$$\sin 40° = \frac{a}{30}$$

$30 \times \sin 40° = a$

$a = 19.3m$ (3 s.f.)

The kite is 19.3m above Siân's head.

1. Calculate the length x in each triangle.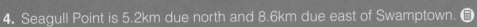

   a)

   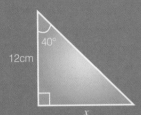

   b)

   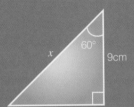

2. Work out the size of the angle θ in each of these triangles.

   a)

   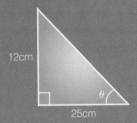

   b)

   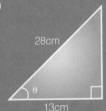

3. A circle has a radius of 10cm. Calculate the length of the chord CD.

   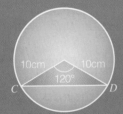

4. Seagull Point is 5.2km due north and 8.6km due east of Swamptown.
   a) Calculate the direct distance from Seagull Point to Swamptown.
   b) Daisy wants to sail directly from Swamptown to Seagull Point.
   On what bearing should she sail?

   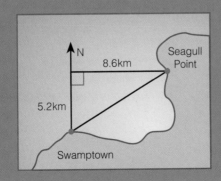

Progress Check

117

Worked questions

1. K, L, M and N are ports on the Blue Sea. Cruise liner C sails 170km due East from L to M, then 150km due North from M to N and finally 40km due West to K. Cruise liner V sails from L to K on a bearing of 55° and cruise liner E sails from M to K on a bearing of 345°.

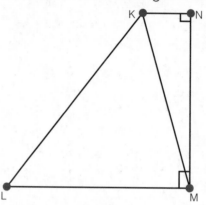

Your first task is to put the above information on to the sketch. Write on the distances and draw the North lines from the required points. Then add the bearings given.

a) From K, V sails to M. What bearing should the captain set? *(3 marks)*

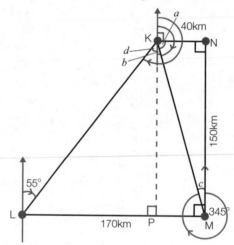

The two angles forming V's new bearing from K can be found either by adding *a* to the right angle (90° + *a*) or by subtracting *b* from 180° (180° − *b*) (angles on a straight line).

To find either *a* or *b*, you first need to find *c*.

c = 360° − 345° = 15°

a therefore is 180° − (90° + 15°) = 75° (angles in a triangle)

Always draw your North lines clearly so that you can easily see parallel lines.

b = *c* = 15° (alternate angles created by the two North lines which are parallel)

V's bearing will be 165°.

b) From K, E sails to L. What bearing should the captain set? *(2 marks)*

E's bearing is 90° + *a* + *b* + *d*

Alternate angles again created by two North lines.

= 90° + 75° + 15° + 55° = 235°

E's bearing will be 235°.

c) How far did V and E sail? *(3 marks)*

For this, you need to calculate the lengths of LK and KM.

Using Pythagoras, KM² = 150² + 40² = 22 500 + 1600 = 24 100

KM = $\sqrt{24\,100}$ = 155km to the nearest km.

KP = MN = 150km

LP = 170 − PM = 170 − 40 = 130km

Using Pythagoras, LK^2 = 150^2 + 130^2 = 22500 + 16900 = 39400

LK = $\sqrt{39400}$ = 198km to the nearest km.

V and E sailed 353km.

2. Katie's dog has fallen on to a ledge on a cliff. The angle of elevation from point Z to the top of the cliff, 70 metres along the beach, is 35°. The angle of elevation to the ledge from Z is 30°.

How far down the cliff from the top does the rescue team need to climb? Round your figures to 1 decimal place. *(3 marks)*

Label the points on the cliff A, B and C. The lines involved are adjacent and opposite to the known angle.

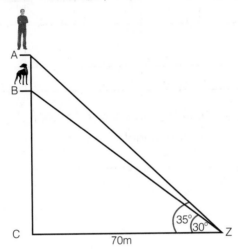

$\tan 30° = \dfrac{BC}{70}$

$\tan 30° \times 70 = BC$

$\tan 30° \times 70 = 40.4m$ rounded to 1 d.p.

$\tan 35° = \dfrac{AC}{70}$

$\tan 35° \times 70 = AC$

$\tan 35° \times 70 = 49m$

The rescue team will need to climb down 8.6m.

tan 30° = $\dfrac{\text{opposite}}{\text{adjacent}}$

tan 35° = $\dfrac{\text{opposite}}{\text{adjacent}}$

3. The scale on a map is 1: 30000. Two villages are 27cm apart on the map.

What is the real distance between them in kilometres? *(2 marks)*

27 × 30000 = 810000cm

810000cm ÷ 100 = 8100m

8100 ÷ 1000 = 8.1km

The scale 1: 30000 means that 1cm on the map equals 30000cm in reality. 27cm on the map therefore equals 27 × 30000 in real life.

100cm = 1m so to change centimetres into metres we divide by 100.

1000m = 1km so to change metres into kilometres we divide by 1000.

Practice questions

1. Complete this table.

(4 marks)

Name	Number of sides	Number of angles	Lines of symmetry	Order of rotation
	3	3	3	3
Kite		4		
	4	4	0	1
Regular heptagon				
Parallelogram	4	4		2

2. Complete this table by inserting ticks to show the type of each triangle. You may need to tick more than one box for one or more of the triangles.

(3 marks)

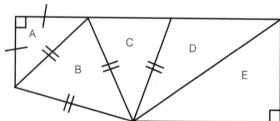

	Equilateral	Isosceles	Right-angled	Scalene
A				
B				
C				
D				
E				

3. Which of the following triangles are congruent?

Which condition meets your answer?

(1 mark)

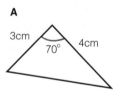

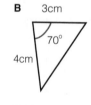

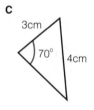

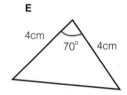

4. What do the following lines on the diagram below represent? *(3 marks)*

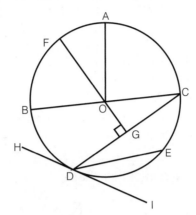

AO

DE

FG

HDI

CA (curved)

BC

5. The interior angle of a regular polygon is 108°.

  a) What is the exterior angle? *(1 mark)*

  b) How many sides does the polygon have? *(1 mark)*

  c) What is the name of this polygon? *(1 mark)*

6. Look at the diagrams A, B, C, D below.

A

B

C

D

  a) Which of these reflections is correct? *(1 mark)*

  b) What is wrong with the other three? *(3 marks)*

7. What is the order of rotation of each of these numbers?   *(3 marks)*

    3

    8

    609

    909

    69

8. Draw the front and side elevations of this shape.   *(2 marks)*

    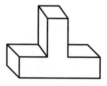

9. Look at the diagram below.

    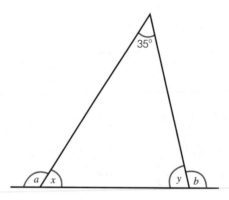

    a) i) What is the sum of the angles a and b?   *(1 mark)*

        ii) Explain your reasoning.   *(2 marks)*

    b) In this diagram, what are the values of x and y?   *(2 marks)*

    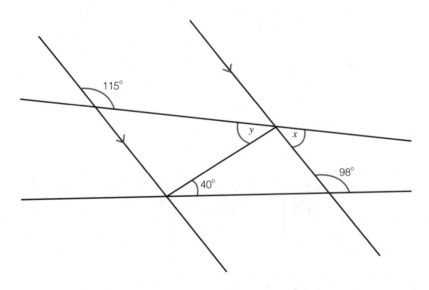

10. The top of the water slide at the Happy Days Theme Park is 26m high. The angle
of depression of the slide from the top is 42°. *(2 marks)*

To have another turn, you have to walk 60m back from B to A, to the bottom of the ramp.

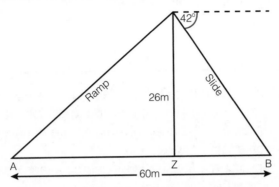

a) What is the length of the slide rounded to the nearest metre? *(1 mark)*

b) What is the length of the ramp? Round any preliminary calculations to the
nearest metre. *(3 marks)*

c) Find the angle of elevation of the ramp from point A rounded to the nearest
whole number. *(1 mark)*

11. A plan of Jack's garden is drawn to a scale of 1 : 200.

Complete the table to show the actual dimensions in metres of Jack's lawn,
patio and flower bed. *(3 marks)*

	Length on plan (cm)	Actual length (m)
Lawn	4cm × 2cm	
Patio	3cm × 2cm	
Flower bed	7cm × 1cm	

12. A, B and C are islands. A cruise ship leaves A for B on a bearing of 135°.

What bearing should the captain set in order to sail from B to C? *(1 mark)*

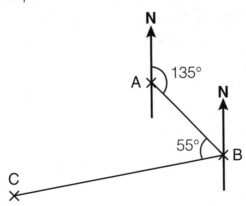

Learning Summary

After studying this chapter you should be able to:

- transform 2-D shapes using translation, reflection and rotation
- enlarge 2-D shapes given a centre of enlargement and a positive scale factor
- transform 2-D shapes by combinations of translations, reflections and rotations
- recognise and use similarity to solve problems
- make constructions and draw loci.

 6.1 # Transformations and similarity

A transformation changes the position or size of a shape. There are four types of transformation: **translation**, **reflection**, **rotation** and **enlargement**.

Reflections and translations

A reflection creates an image of an object on the other side of a mirror line. The mirror line is known as an axis of reflection. The size and shape of the figure are not changed. Understanding reflections and translations helps us to appreciate patterns in nature.

Example

Reflect triangle ABC in:

a) the x-axis and label it D

b) the line $y = -x$ and label it E

c) the line $x = 5$ and label it F.

Triangles D, E and F are congruent to triangle ABC.

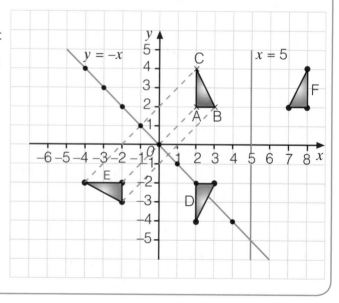

A translation moves objects from one place to another. The size and shape of the object are not changed. **Vectors** are used to describe the distance and direction of the translation. A vector is written as $\begin{pmatrix} a \\ b \end{pmatrix}$

a represents the horizontal movement and b represents the vertical movement.

Examples

1. Draw the image of ABCD after a translation of 4 squares to the left and 3 squares up.

   ABCD and A'B'C'D' are congruent.

2. a) Translate ABC by the vector $\begin{pmatrix} 2 \\ 1 \end{pmatrix}$ and label it P.

   b) Translate ABC by the vector $\begin{pmatrix} -3 \\ -2 \end{pmatrix}$ and label it Q.

   Triangles ABC, P and Q are all congruent.

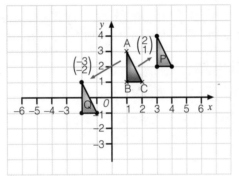

> Equip yourself with squared paper and tracing paper. Draw the two axes and then draw a simple shape like a triangle in one of the quarters. Make reflections in each of the other quarters. Don't just use the x and y-axes. Make other lines your mirror lines. Trace your original shape on tracing paper, cut it out and use it to check that your drawings are correct. Then try the same thing with some translations.

Rotations

Rotations turn an object through an angle about some fixed point. This fixed point is called the centre of rotation. The size or shape of the object is not changed.

To describe a rotation, give three pieces of information:
- The centre of rotation
- The direction of rotation (clockwise or anticlockwise)
- The angle of rotation

For example, this is a 90° rotation about O, in a clockwise direction (also known as a $\frac{1}{4}$ turn clockwise).

By convention, an anticlockwise rotation is positive and a clockwise rotation is negative.

Example

Rotate triangle ABC:

a) 90° clockwise about (0, 0) and label it R.

b) 180° about (0, 0) and label it S.

c) 90° anticlockwise about (−1, 1) and label it T.

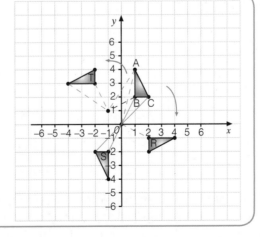

> On squared paper, draw the axes and draw a simple shape in any of the quarters. It is better to use shapes with an odd number of sides. Cut a piece of tracing paper the same size as your graph and trace your shape anywhere around the middle of the tracing paper. Place it directly over the original shape. Place a sharp pencil point on any pair of co-ordinates (it doesn't have to be (0, 0)) and slowly turn the tracing paper around that point. Turn your shape through four directions and see how it changes. Try lots of these until you can do them accurately without the tracing paper.

Enlargements

Enlargements change the size but not the shape of the object. The centre of enlargement is the point from which the enlargement takes place. The **scale factor** (k), sometimes known as the multiplier indicates how many times the length of the original figure has changed. Enlargement with a scale factor k makes the lengths k times longer:

- If the scale factor is greater than 1, the object becomes bigger.
- If the scale factor is less than 1, the object becomes smaller.

The terms multiplication and enlargement are still used even when the scale factor (or multiplier) is less than 1.

A negative scale factor places the image on the opposite side of the centre of enlargement to the object.

When asked to describe an enlargement, you must include the scale factor and the position of the centre of enlargement.

> Multipliers and ratio are used in enlargement and similar figures. See Topic 1.4.

> Draw a triangle on squared paper. Work out the position of the centre of enlargement and choose your scale factor. Complete the enlargement on the squared paper. Make sure you practise decreasing the size, as well as increasing. Next try it the other way round, so that you start by choosing a centre of enlargement. Draw the rays and look at the different sizes of triangles you can draw in.

Example

a) Enlarge triangle ABC by a scale factor of 2, centre (0, 0). Label the image A'B'C'.

b) Enlarge triangle ABC by a scale factor $-\frac{1}{2}$, centre (0, 0). Label the image A"B"C".

> Notice that each side of the enlargement is twice the size of the original. OC' = 2 × OC.

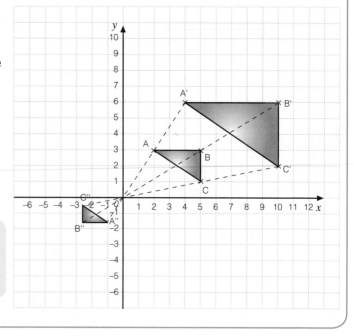

ABC has been enlarged with a scale factor $\frac{1}{2}$ to give A'B'C'. The centre of enlargement is at O.

The length of OA' is $\frac{1}{2}$ OA.

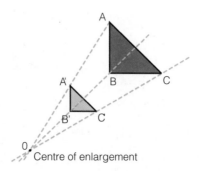

Centre of enlargement

Two successive enlargements with scale factors k_1 and k_2 are equivalent to a single enlargement with scale factor $k_1 \times k_2$.

Combining transformations

Example

a) Reflect ABC in the x-axis.
Label the image $A_1B_1C_1$.

b) Reflect $A_1B_1C_1$ in the y-axis.
Label the image $A_2B_2C_2$.

The single transformation that
maps ABC directly onto $A_2B_2C_2$
is a rotation of 180°, centre
(0, 0).

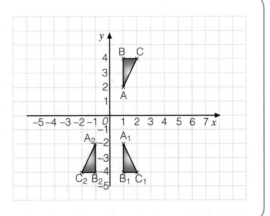

Similar figures

Similar figures are the same shape but different sizes. This means that
shapes that have been enlarged are similar. Corresponding angles are equal.
Corresponding lengths are in the same ratio.

The corresponding angles of these two triangles are equal, which makes
them similar:

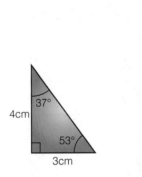 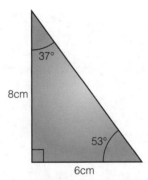

The missing lengths of similar figures can be found because corresponding
lengths are in the same ratio:

 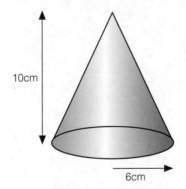

The radius of the base of the first cone is 3cm and the radius of the base of
the second cone is 6cm, so the ratio of the bases is 3 : 6 = 1 : 2

So the scale factor for the enlargement is 2.

Therefore the height of the larger cone will be 5 × 2 = 10cm.

Example

Find the missing length a.

Method 1

The ratio of the bases is 12 : 9

This simplifies to 1 : 0.75 ← Divide by 12 to get the ratio in the form 1 : n

The two shapes are similar (because the corresponding angles are equal) so the ratio of the other side will also be 1 : 0.75

The scale factor of the enlargement is 0.75

The corresponding length of the smaller triangle a, is $0.75 \times 11 = 8.25$cm

Method 2

Set up a fraction with the corresponding sides:

$\dfrac{a}{11} = \dfrac{9}{12}$ ← Corresponding sides are in the same ratio.

$a = \dfrac{9}{12} \times 11$ ← Multiply both sides by 11.

$a = 8.25$cm

Progress Check

1. On the diagram:
   a) Translate ABC by the vector $\begin{pmatrix} -3 \\ 1 \end{pmatrix}$.
   Label it P.
   b) Reflect ABC in the line $y = x$.
   Label it Q.
   c) Reflect ABC in the line $y = -1$.
   Label it R.
   d) Rotate ABC 180° about (0, 0).
   Label it S.

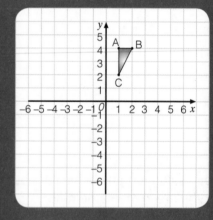

2. Enlarge shape P with a scale factor of 3.

3. The two cylinders are similar.

   Work out the value of x.

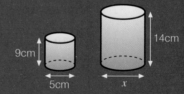

6.2 Constructions

 17

Constructions

You need to know how to construct the following, using a pair of compasses and a ruler. It often helps to sketch a diagram first.

Triangle To construct a triangle: • Draw the longest side. • With the compass point at A, draw an arc of radius 4 cm. • With the compass point at B, draw an arc of radius 5cm. • Join A and B to the point where the two arcs meet. 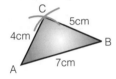	**The perpendicular bisector of a line** • Draw a line XY • Draw two arcs with the compass, using X as the centre. The compass must be set at a radius greater than half the distance of XY. • Draw two more arcs with Y as the centre (keep the compass the same distance apart as before). • Join the two points where the arcs cross. • AB is the perpendicular bisector of XY. • N is the midpoint of XY.
The perpendicular from a point to a line • From P draw arcs to cut the line at A and B. • From A and B draw arcs with the same radius to intersect at C. • Join P to C; this line is perpendicular to the line AB.	**The perpendicular from a point on a straight line** • With a compass set to a radius of several cm, and centred on N, draw arcs to cut the line at A and B. • Construct the perpendicular bisector of the line segment AB as shown above.

To bisect an angle

• Draw two lines XY and YZ to meet at an angle.
• Using a pair of compasses, place the point at Y and draw two arcs on XY and YZ.
• Place the compass point at the two arcs on XY and YZ and draw arcs to cross at N. Join Y to N. YN is the bisector of angle XYZ.

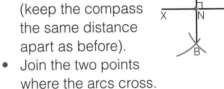

Practise drawing all these constructions. Do not rub out your construction lines when drawn. You need to feel at ease using your compass so that you can be accurate and confident.

Loci

The **locus** of a point is the set of all the possible positions which that point can occupy, subject to some given conditions or rules. The plural of locus is loci.

The locus of the points that are a constant distance from a fixed point is a circle.	The locus of the points that are equidistant from two lines is the line that **bisects** the angle between the lines. 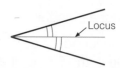
The locus of the points that are equidistant from two points X and Y is the perpendicular bisector of XY. 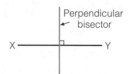	The locus of the points that are a constant distance from a line is a pair of parallel lines above and below the line. (Remember that a line is infinitely long.) 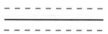 For a line segment, there would be semicircles on either end. 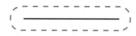

Example

The diagram shows a rectangular field. Gertie the goat can only eat grass in the area that satisfies these given conditions:

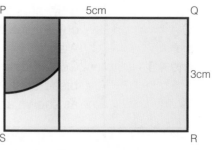

(Scale 1cm = 1m)

- Not more than 2m from P

- At least 3.5m away from the wall QR.

The green-shaded region shows where Gertie can eat the grass.

Scale diagrams are used when showing the locus of points. See Topic 5.2.

1. A gold coin is buried in a rectangular field. It is 4m from T and equidistant from RU and RS. Mark with an X the position of the gold coin.

2. Draw an angle of 40°. Bisect the angle accurately, showing all construction lines.

3. Construct the perpendicular bisector of a 10cm line AB.

4. Emily is redesigning her garden. She wishes to plant a tree in the garden. The tree must be at least 4m from the house and at least 10m from the centre of the pond.

   Show accurately the region in which the tree can be planted.

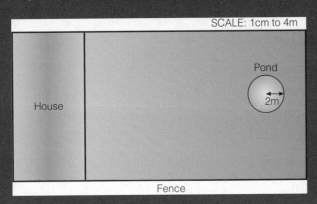

Progress Check

Worked questions

Opposite angles created where two lines cross are equal. The bases of the triangles are parallel, creating alternate angles, that are equal. The first job when dealing with similar triangles is to make sure that you match the sides and angles correctly. All sorts of tricks are employed to divert you from the correct choice. Sometimes it is a good idea to draw the triangles separately, with each in the same orientation.

1. These triangles are similar.   *(2 marks)*

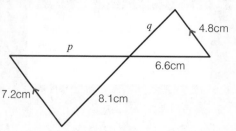

Find the lengths marked p and q.

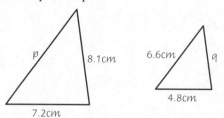

Looking at the two bases:

Next you have to determine the scale factor.

$Scale\ factor = \dfrac{7.2}{4.8} = 1.5$

Now make equations for the unknown sides.

$\dfrac{p}{6.6} = 1.5$

$p = 1.5 \times 6.6 = 9.9cm$

$\dfrac{8.1}{q} = 1.5$

$q = \dfrac{8.1}{1.5} = 5.4cm$

$p = 9.9cm\ q = 5.4cm$

2. Look at this diagram:

First of all, establish the scale factor. The vertical length has increased from 2 to 4. Because the triangle has increased in size the scale factor will be a whole number.

The horizontal length has increased from 3 to 6.

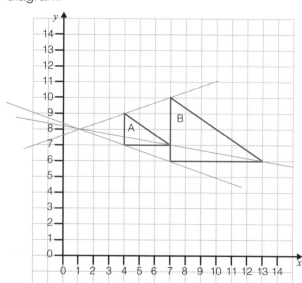

a) Describe the transformation which takes triangle A to B.   *(2 marks)*

4 ÷ 2 = 2

6 ÷ 3 = 2

The scale factor is 2.

Remember, enlargements must always have two details – the scale factor and the centre of enlargement. To find the centre of enlargement, draw lines from the three corresponding points of the triangles in a direction in which they are going to converge. The point at which they converge is the centre of enlargement.

The centre of enlargement is (1, 8).

The transformation which takes triangle A to B is enlargement around the centre (1, 8) with a scale factor of 2.

The word 'enlargement' is used even when the shape is going to decrease in size. Place a small dot on the centre of enlargement. Draw lines from each of the points of B to (3, 2). Extend your lines beyond the centre of enlargement and the points of the triangle.

b) Enlarge triangle B using a scale factor of $\frac{1}{4}$ around (3, 2). Label it C.   *(3 marks)*

4 ÷ 4 = 1

6 ÷ 4 = $1\frac{1}{2}$

Now work out the lengths of the new triangle.

Now find the point at which those new lengths fit between the ray lines. Between (4, 4) and (4, 3) is a dimension of 1, which fits perfectly between the lines coming from points R and S. Between (4, 3) and (5, 5$\frac{1}{2}$) is a dimension of $1\frac{1}{2}$, which fits perfectly between the lines coming from points S and T. Draw your new triangle and don't forget to label it C.

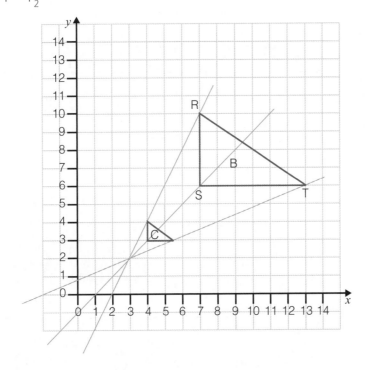

The scale of the drawing is 1cm = 2m. To show the area that Steve can build on without needing planning permission, lines are needed 1cm away from the perimeter on the plan.

Using a ruler, carefully measure 1cm along the perimeter each way from each corner. Mark these points with small dots. To ensure that the angles of your new lines will be at right angles, set your compass to 1cm and draw arcs from each of the dots. Place a dot where the two arcs cross and carefully draw the lines.

A circle needs to be drawn around the pond to show the area that Steve wishes to leave clear. Set your compass to 2.5cm and draw the new circle.

In the same way, a semi-circle needs to be drawn around the flower bed to show the area that Steve's wife wants left clear. The radius of this semi-circle will be 3m + radius of flower bed. The diameter of the flower bed is 7m = a radius of 3.5m. Set your compass to 3.25cm and draw the new semi-circle.

By measuring, you will be able to ascertain the area where the garage can be built. Shade this area. Remember too – never rub out your construction lines.

3. Steve wants to build a new garage without planning permission.

Provided that his new garage is 2 metres away from his perimeter fencing, he does not need planning permission. Steve does not want his garage too close to his garden pond, so he wants to leave at least 3 metres between the garage and the pond. Steve's wife does not want the garage to overshadow her flower bed, so she wants at least 3 metres between the flower bed and the garage. The dimensions of the garage are 7 metres by 4 metres. Shade the area in which Steve can build his garage.

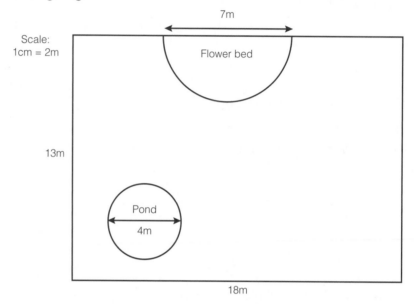

(4 marks)

The radius of this circle will be 3m + radius of pond. The diameter of the pond is 4 metres = a radius of 2m. So the new radius = 3m + 2m = 5m = 2.5cm.

New radius = 3.5m + 3m = 6.5m = 3.25cm

Dimensions of garage = 7m × 4m

$$= 3.5cm × 2cm \text{ on the plan}$$

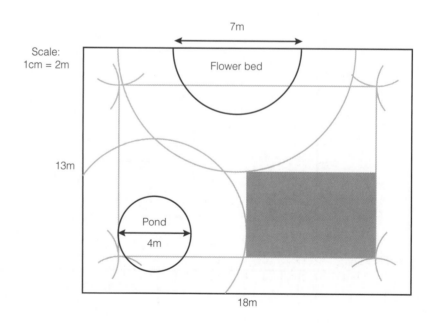

Practice questions

1. Look at this diagram:

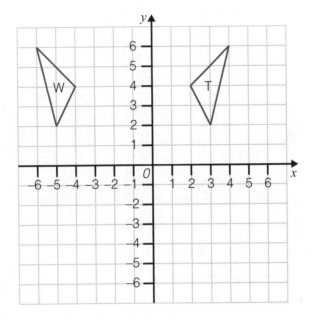

a) What is the mirror line for the reflection on to the triangle T at W?   *(1 mark)*

b) Draw a triangle which is a reflection of the triangle T in the *x*-axis.
    Label the new triangle S.   *(1 mark)*

2. Look at this diagram:

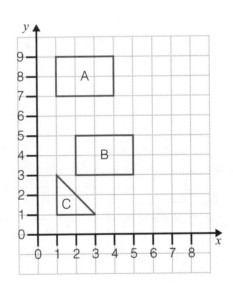

a) What transformation takes rectangle A to B?   *(2 marks)*

b) Translate triangle C by vector $\begin{pmatrix} 5 \\ 5 \end{pmatrix}$ and label it D.   *(1 mark)*

c) If C were translated by vector $\begin{pmatrix} -6 \\ 4 \end{pmatrix}$, what vector will translate it back
    to its original position?   *(1 mark)*

135

3. Look at this diagram:

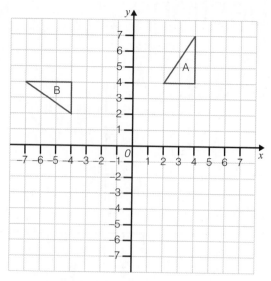

a) Describe the rotation that takes triangle A onto B. *(2 marks)*

b) Rotate triangle A half a turn around (1, 1). Label it C. *(2 marks)*

4. Ismail's favourite photograph of his grandson measures 8cm × 6cm. He wants to enlarge the length to 12cm.

a) What scale factor should he use? *(1 mark)*

b) What will be the new width? *(1 mark)*

5. Look at this diagram:

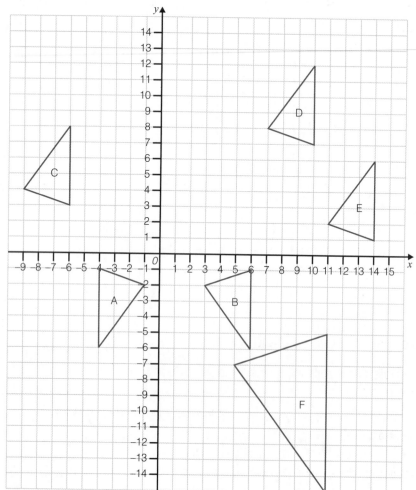

   a) Which transformation takes triangle B to A?   *(2 marks)*

   b) Describe the transformation which takes A to C.   *(3 marks)*

   c) Describe the transformation of E to D.   *(2 marks)*

   d) Which transformation takes F to B?   *(3 marks)*

6.   These triangles are similar.

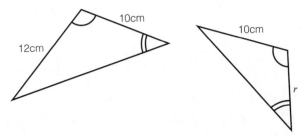

   Find the length of the side marked r rounded to one decimal place.   *(1 mark)*

7.   Under this staircase is a cupboard. At its highest point it is 3m high and it goes back under the stairs for 2m.

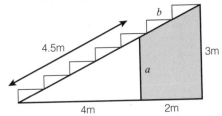

   Using similar triangles, find the two remaining dimensions a and b.   *(2 marks)*

8.   Mobile phone service on Geometry Island is covered by three masts:

   The mast positioned on the West has a range of 150km in all directions.

   The mast positioned on the East has a range of 250km in all directions.

   The mast positioned in the centre has a range of 300km in all directions.

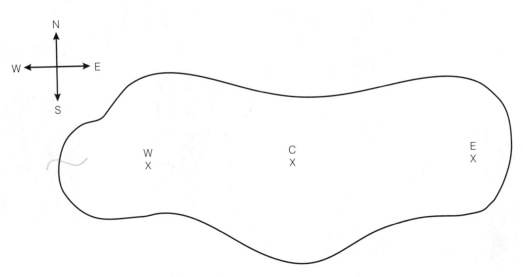

   Shade the parts of the island which have no service. Use a scale of 1cm = 100km.   *(3 marks)*

Learning Summary

After studying this chapter you should be able to:
- interpret numbers on a range of measuring instruments, and choose and use appropriate units
- use metric and imperial units, and solve problems involving the conversion of units
- find the area and perimeter of 2-D shapes
- know and use the formulae for the circumference and area of a circle
- calculate the volume of a variety of 3-D solids.

18 7.1 Units of measurement

Estimating

Estimating is a useful skill in everyday life. You need to be able to estimate measures such as:

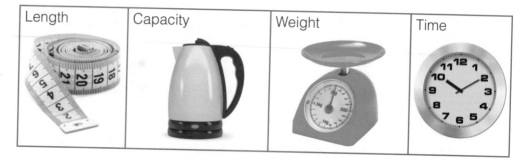

Length	Capacity	Weight	Time

A door is about 2m high.	A can of soft drink holds about 330ml or $\frac{1}{2}$ pint.	A bag of sugar holds 1kg or about 2.2lb.

Have an estimating day where you record your estimates of different objects you come across. Then check your estimates by measuring the objects.

Metric units

Metric units include kilometres (km), metres (m), kilograms (kg) and litres (l).

Length	Weight	Capacity
10mm = 1cm	1000mg = 1g	1000ml = 1 litre
100cm = 1m	1000g = 1kg	100cl = 1 litre
1000m = 1km	1000kg = 1 tonne	1000cm³ = 1 litre

When converting one unit to another, try to decide first whether your answer will be larger or smaller, then multiply or divide as appropriate:

Rule	Examples
If changing from small units to large units, you divide	500cm = 5m (÷ 100)
	3500g = 3.5kg (÷ 1000)
If changing from large units to small units, you multiply	5 litres = 500cl (× 100)
	25cm = 250mm (× 10)

These diagrams may help you to remember:

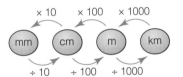 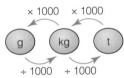 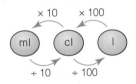

Imperial units

Imperial units include miles, yards, stones, pints, etc. They are sometimes thought of as the 'old fashioned' units of measurement.

Length	Weight	Capacity
1 foot = 12 inches	1 stone = 14 pounds (lb)	20 fluid oz = 1 pint
1 yard = 3 feet	1 pound = 16 ounces (oz)	8 pints = 1 gallon

Here are some approximate comparisons between metric and imperial units:

Length	Weight	Capacity
2.5cm ≈ 1 inch	25g ≈ 1 ounce	1 litre ≈ $1\frac{3}{4}$ pints
30cm ≈ 1 foot	1kg ≈ 2.2 pounds	4.5 litres ≈ 1 gallon
1m ≈ 39 inches		
8km ≈ 5 miles		

You must know these relationships inside out. Make three cards. On one of them write 10, on the second 100 and on the third 1000. On the other side write the relationships which match. Just write the units – no numbers. So, on the back of 100 you would write 'm and cm' and 'cl and litre'. Next, turn the cards face up so the numbers are showing. Look at them and recite the facts which are on the back. You can also use the cards with the units face up and recite what you have to do to change one into the other. When you look at mm and cm, for instance, you would say 'mm to cm – divide by 10'.

Being able to convert between different measurements is necessary when converting scales on maps and scale diagrams. See Topic 5.2.

Look at tins, bottles and packages in the kitchen to see the units in which they are measured. Check with an adult before you do so.

Most measuring jugs and kitchen scales show both metric and imperial units. Ask an adult first, but by using these jugs and scales you can reinforce your understanding of conversions. For example, fill a measuring jug with different amounts of water. Read and write down both the metric and the imperial measurements for different volumes of water.

Examples

1. Change 25km into miles.

    8km ≈ 5 miles so 1km ≈ $\frac{5}{8}$ mile.

    25km ≈ 25 × $\frac{5}{8}$ = 15.6 miles (1 d.p.)

2. A plate is 6 inches across. Roughly how many centimetres is this?

    2.5cm ≈ 1 inch   6 inches ≈ 6 × 2.5 = 15cm

Choosing the correct units of measurement

When measuring, it is important that sensible units are used.

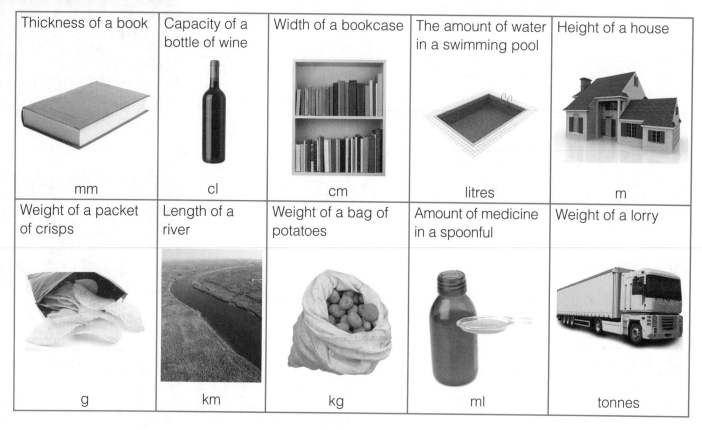

Thickness of a book	Capacity of a bottle of wine	Width of a bookcase	The amount of water in a swimming pool	Height of a house
mm	cl	cm	litres	m
Weight of a packet of crisps	Length of a river	Weight of a bag of potatoes	Amount of medicine in a spoonful	Weight of a lorry
g	km	kg	ml	tonnes

Time measurement

Units of time

You should know these units of time.

There are 60 seconds in one minute.

There are 60 minutes in one hour.

There are 24 hours in one day.

There are seven days in one week.

There are 52 weeks in one year.

There are 365 days in a year, or 366 in a leap year.

This clock reads 10 past 7. The short hand tells us the hour, and the long hand tells us the minutes.

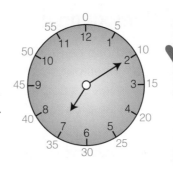

The 12 and 24-hour clock

Time can be measured using the 12- or 24-hour clock.

The 12-hour clock uses am and pm. Times before midday are am; times after midday are pm.

The 24-hour clock numbers the hours from 0 to 24. Times are written with four figures.

12-hour clock		24-hour clock
2.42pm	is the same as	1442
3.30am	is the same as	0330

24-hour clock		12-hour clock
1527	is the same as	3.27pm
0704	is the same as	7.04am

Timetables

24-hour clock times often appear on bus and train timetables.

> You need a clock with a face and a digital clock, both set to the same time. At different times of the day look at them and make sure you understand the readings. Have a go at looking at one with the other covered. Say what the time will be on the one which is covered. If the digital clock doesn't give 24-hour readings, practise converting to the 24-hour clock as well.

Example

The train timetable gives the times from London to Manchester.

London Euston	0702	0740	Every 60 minutes until	1100	1400
Watford Junction	0732	0812		1130	1430
Stoke-on-Trent	0850	0930		–	1545
Manchester Piccadilly	0940	1015		1315	1640

The 0850 train from Stoke-on-Trent.

The 0740 train from London Euston arrives at 1015.

The 1100 train from London Euston does not stop at Stoke-on-Trent.

a) Diana is travelling from Watford Junction to Manchester Piccadilly. If she catches the 0732 train from Watford, how long is her journey?

Departs Watford: 0732   Arrives Manchester: 0940
Time = 2 hours 8 minutes

b) Sebastian arrives at London Euston at 1242. How long does he have to wait for the next train to Manchester?

1242 → 1300 = 18 minutes
1300 → 1400 = 1 hour
Waiting time is 1 hour 18 minutes

> Being able to read from a variety of tables is important when interpreting data. See Topic 9.3.

> Collect a bus timetable and a train timetable. Study the timetables to make sure you understand them properly. You could ask an adult to set you questions about the times shown.

Compound measures

Speed can be measured in kilometres per hour (km/h), miles per hour (mph) and metres per second (m/s). These are all compound measures because they involve a combination of two basic measures. **Density** and pressure are also examples of compound measures.

The abbreviation for 'per' is a 'p' or '/' and is used to mean 'for every' or 'in every', e.g. mph (miles travelled in every hour).

$$\text{Average speed} = \frac{\text{total distance travelled}}{\text{total time taken}}$$

$$\text{Average speed} = \frac{d}{t}$$

From this speed formula two others can be obtained:

$$\text{Time} = \frac{\text{distance}}{\text{speed}}$$

$$\text{Distance} = \text{speed} \times \text{time}$$

> Being able to substitute and change the subject of a formula is important when working with compound measures. See Topic 3.1.

> Write the letters s (for speed), t (for time) and d (for distance) on three separate cards. Write the symbols $\times$ and $\div$ on two separate cards. Mix up the cards and rearrange them correctly so that s is the subject of the formula. Then try again, making t and d the subject of the formula.

> Multiplying by decimals is useful in these calculations. See Topic 2.1.

Examples

1. Lynette walks 10km in 4 hours. Find her average speed.

$$S = \frac{d}{t}$$

$$S = \frac{10}{4}$$

$$S = 2.5\text{km/h}$$

2. Mr Rosenthal drove a distance of 250 miles at an average speed of 70 miles per hour. How long did the journey take?

$$t = \frac{d}{s}$$

$$t = \frac{250}{70}$$

$$t = 3.57 \text{ hours}$$

3.57 hours must be changed into hours and minutes. To do this:
- subtract the hours
- multiply the decimal part by 60, i.e. 0.57... × 60 = 34 minutes (nearest minute)

> We multiply by 60 because there are 60 minutes in 1 hour.

Journey time = 3 hours 34 minutes

To calculate density, volume and mass, use:

Density = $\dfrac{\text{mass}}{\text{volume}}$

$$D = \dfrac{M}{V}$$

Volume = $\dfrac{\text{mass}}{\text{density}}$

Mass = density × volume

Example

Find the density of an object whose mass is 600g and whose volume is 50cm³.

Density = $\dfrac{M}{V}$

$= \dfrac{600}{50}$

$= 12\text{g/cm}^3$

To calculate pressure use:

Pressure = $\dfrac{\text{force on surface}}{\text{surface area}}$

$$P = \dfrac{f}{a}$$

1. Approximately how many pounds are in 2kg of sugar?

2. The following information shows how long it takes to fly between some cities.

From	To	Time
London	Paris	1 hour 15 minutes
London	Tokyo	12 hours 40 minutes
Paris	Sydney	22 hours 10 minutes
Tokyo	Sydney	9 hours 15 minutes

Progress Check

a) Bina flies from London to Paris and then from Paris to Sydney. How long is the flight time in total?

b) Jessica leaves Tokyo at 1800. What time will it be in Tokyo when she is due to land in Sydney?

3. Change 5 litres into ml.

4. Change 5 inches into cm.

5. The mass of an object is 500g. If its density is 6.2g/cm³, what is the volume of the object? 🔲

143

7.2 Area and perimeter of 2-D shapes

Estimating areas of 2-D shapes

The distance around the outside edge of a shape is called the **perimeter**. The **area** of a 2-D shape is the amount of space it covers. Units of area are mm², cm² and m².

Areas of irregular shapes can be found by counting the squares the shape covers. Label the squares as you count them. Try to match up parts of squares to make a whole one.

Examples

1. Find the area of this shape.

   This shape has an area of about 20.5 square units.

1	2	3	4
5	6	7	8
9	10	11	12
13	14	15	16
17	18	19	

   These make one whole square

   This is half a square

2. Find the perimeter of this shape.

   Perimeter = 4 + 5 + 3 + 2.7 + 2.7
   = 17.4cm

Areas of quadrilaterals and triangles

You need to know the formulae for these 2-D shapes.

Area of a rectangle	**Area of a parallelogram**
Area = length × width	Area = base × perpendicular height
$A = l \times w$	$A = b \times h$
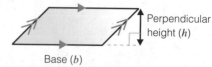	

Area of a triangle	**Area of a trapezium**
Area = $\frac{1}{2}$ × base × perpendicular height	Area = $\frac{1}{2}$ × (sum of parallel sides) × (perpendicular height between sides)
$A = \frac{1}{2} \times b \times h$	$A = \frac{1}{2} \times (a + b) \times h$
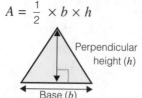	

Make a set of revision cards with the name and a picture of each of the shapes mentioned here. On the back write the formula you need to calculate that shape's area. Practise looking at the name of the shape and reciting the formula for its area.

Examples

1. Find the area of these shapes, giving your answers to 3 s.f. where necessary.

a)

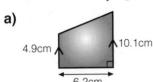

$A = \frac{1}{2} \times (a + b) \times h$

$\quad = \frac{1}{2} \times (4.9 + 10.1) \times 6.2$

$\quad = 46.5\text{cm}^2$

b)

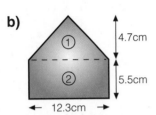

Split the shape into two parts and find the area of each.

Area of 1 $= \frac{1}{2} \times b \times h$

$\quad = \frac{1}{2} \times 12.3 \times 4.7$

$\quad = 28.905\text{cm}^2$

Area of 2 $= b \times h$

$\quad = 12.3 \times 5.5$

$\quad = 67.65\text{cm}^2$

Total area $= 1 + 2$

$\quad = 28.905 + 67.65$

$\quad = 96.555$

$\quad = 96.6\text{cm}^2$ (3 s.f.)

2. If the area of this triangle is 55cm², find the height giving your answer to 3 s.f.

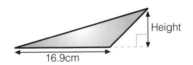

$A = \frac{1}{2} \times b \times h$

$55 = \frac{1}{2} \times 16.9 \times h$ Substitute values into the formula.

$55 = 8.45 \times h$

$h = \dfrac{55}{8.45}$ Divide both sides by 8.45

$h = 6.51\text{cm}$ (3 s.f.)   Notice that rounding does not take place until the end.

3. The perimeter of this shape is 30cm. Work out the length of the longest side of this shape.

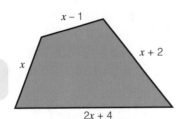

$x + x - 1 + 2x + 4 + x + 2 = 30$ Set up an equation first.

$5x + 5 = 30$ Collect like terms.

$5x = 30 - 5$

$5x = 25$

$x = \dfrac{25}{5}$ Divide both sides by 5.

$x = 5\text{cm}$

Length of the longest side $= 2x + 4$

$\quad = 2 \times 5 + 4$

$\quad = 14\text{cm}$

Writing equations, solving equations, substitution into formulae and rearranging formulae all apply to questions involving finding areas of shapes. See Topics 3.1 and 3.2.

Circumference and area of a circle

It is important that you remember these formulae:

Circumference = π × diameter	$C = π × d$
= 2 × π × radius	$C = 2 × π × r$
Area = π × (radius)²	$A = π × r^2$

Examples

1. Find the circumference and area of this circle.

   Use π = 3.142
   $C = π × d$
   $C = 3.142 × 10$
   $C = 31.42$cm

   $A = π × r^2$
   $A = 3.142 × 5^2$
   $A = 78.55$cm²

   > Halve the diameter to obtain the radius, which we need when finding the area.

   10cm

2. A circular fish pond has a circumference of 12m. Work out the length of the diameter to 1 d.p. Use π = 3.142

   $C = π × d$
   $12 = 3.142 × d$
   $\dfrac{12}{3.142} = d$
   So $d = 3.8192$
   $d = 3.8$m (1 d.p.)

Being able to change the subject is important in this type of question. See Topic 3.1.

3. A garden is shown. Fencing is to be placed around the perimeter of the garden. Work out the length of fencing that is needed. Use π = 3.142. Give your answer to 2 s.f.

   8m

   Work out the circumference of the semicircle first:
   $C = π × d$
   $C = 3.142 × 8$
   $C = 25.136$

   > We need to halve this to find the circumference of the semicircle.

   $\dfrac{25.136}{2} = 12.568$m

   Adding the horizontal distance will give the perimeter:

   $12.536 + 8 = 20.568$

   The length of fencing = 20.568m
   = 21m (2 s.f.)

When finding an **arc** length or a **sector** area of a circle, it is important to remember that they are just a fraction of the circumference or the area of the circle.

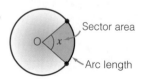

$$\text{Arc length} = \frac{x}{360°} \times \pi \times d$$

$$\text{Sector area} = \frac{x}{360°} \times \pi \times r^2$$

where x is the size of the angle between the two bounding radii.

You need to be familiar with the π button on your calculator. See Topic 2.1.

Areas of enlarged shapes

A common mistake is to assume that an enlargement with scale factor 3 makes the area 3 times larger. In fact, the area of the image is 9 times the area of the original shape.

If a shape is enlarged by a scale factor k, then the area of the enlarged shape is k^2 times bigger.

For example, if $k = 2$:

- the length of the enlarged shape is twice as big
- the area of the enlarged shape becomes four times as big (i.e. $2^2 = 4$).

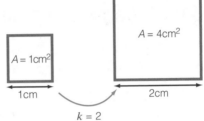

Scale factors are also used in ratio, proportion and similarity problems. See Topics 1.4 and 6.1.

Changing area units

You should know that $1m^2 = 10\,000cm^2$

For example, this square has a length of 1 metre. This is the same as a length of 100cm.

Hence $1m^2 = 100 \times 100cm^2$

$1m^2 = 10\,000cm^2$

Always check that the measurements are in the same units before you calculate an area.

1. Work out the areas of the following shapes, giving your answers to 3 s.f. Use the π key on your calculator. ▣

a)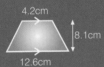

b)

c)

d)

2. Work out the area of the region shaded green. Use the π key on your calculator. ▣

Progress Check

3. Change $5m^2$ to cm^2. Which answer is correct?
   A $5000cm^2$ **B** $500cm^2$ **C** $50\,000cm^2$ **D** $500\,000cm^2$

4. The diagram shows the plan of a room. Carpet is being laid on the floor. $1m^2$ of carpet costs £35.99.
   Work out the total cost of carpeting the room. ▣

7.3 Volume of 3-D solids

The **volume** of a 3-D solid is the amount of space it occupies. Units of volume are cubic millimetre (mm³), cubic centimetre (cm³) and cubic metre (m³).

The volume of a 3-D solid can be found by counting the number of 1cm³ cubes.

For example, the volume of the solid opposite is 24cm³. Each cube has a volume of 1cm³ (1 cubic centimetre).

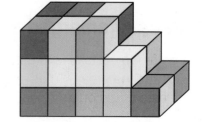

Calculating volume and surface area

> By knowing the properties of 3-D shapes, you can work out which are prisms. See Topic 5.1.

A **prism** is any solid which can be cut into slices that are all the same shape. A prism has a uniform cross-section.

Volume of a cuboid

Volume = length × width × height

$$V = l \times w \times h$$

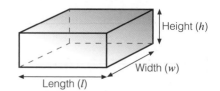

> To find the volume of prisms you need to know how to find the area of the end face, i.e. the area of cross-section. See Topic 7.2

Volume of a prism

Volume = area of cross-section × length

$$V = A \times l$$

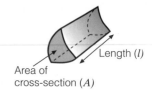

Surface area of a cuboid

Surface area = the sum of the areas of the faces

$$= 2 \times (l \times h) + 2 \times (w \times h) + 2 \times (w \times l)$$

Two faces have area $l \times h$

Two faces have area $w \times h$

Two faces have area $w \times l$

> Collect some small cartons of different sizes. Open them up and measure them to calculate the surface area. Draw the net each one makes and write the measurements on your drawing.

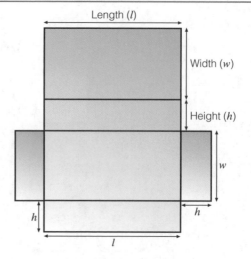

Examples

1. Look at the box opposite.

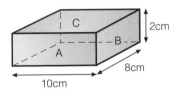

a) Work out the volume of the box.

Volume = $l \times w \times h$

$= 10 \times 8 \times 2$

$= 160cm^3$

b) The box contains a present. Work out the amount of paper needed to wrap the box, assuming there is no overlap.

Surface area = 2 × (face A + face B + face C)

Face A = 10 × 2 = 20cm²

Face B = 8 × 2 = 16cm²

Face C = 10 × 8 = 80cm²

Total amount of paper needed = 2 × (20 + 16 + 80)

$= 2 \times 116$

$= 232cm^2$

> Drawing the nets of 3-D shapes helps work out the surface area of the shapes. See Topic 5.1.

2. The cross-section of a solid is in the shape of a trapezium. Work out the volume of the solid.

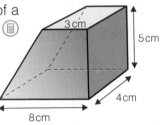

Area of cross-section:

$A = \frac{1}{2} \times (a + b) \times h$

$A = \frac{1}{2} \times (3 + 8) \times 5 = 27.5cm^2$

Volume = 27.5 × 4

$= 110cm^3$

Volume and surface area of a cylinder

Cylinders are prisms whose cross-section is a circle.

Volume = area of cross-section × length

$V = \pi \times r^2 \times h$

> $\pi \times r^2$ is the area of the circle at the end of the cylinder.

Surface area of a cylinder = $2\pi r^2 + 2\pi rh$

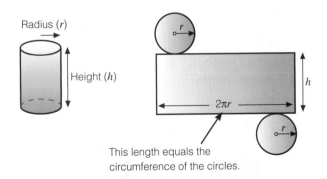

Radius (r)

Height (h)

$2\pi r$

This length equals the circumference of the circles.

Example

Cat food is sold in tins. Work out the following using $\pi = 3.142$. Remember to put units in your answer.

a) The volume of cat food that the tin contains.

$V = \pi \times r^2 \times h$
$V = 3.142 \times 4^2 \times 10$
$V = 502.72$
$V = 503\text{cm}^3$ (3 s.f.)

b) The total area of metal needed to make the tin.

Total area of metal $= 2\pi r^2 + 2\pi rh$
$A = 2 \times 3.142 \times 4^2 + 2 \times 3.142 \times 4 \times 10$
$A = 100.544 + 251.36$
$A = 351.904$
$A = 352\text{cm}^2$ (3 s.f.)

Volumes of enlarged solids

If a solid is enlarged by a scale factor k, the volume of the enlarged solid is k^3 times bigger.

For example, if a cube of length 1cm is enlarged by a scale factor of 2, the volume of the enlarged cube is eight times bigger ($2^3 = 8$).

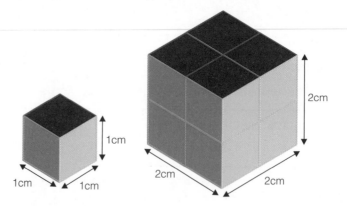

Converting volume units

It is better to change all the lengths to the same unit before starting a question.

For example, this cube has a length of 1m.
This is the same as a length of 100cm.

Hence $1\text{m}^3 = 100 \times 100 \times 100\text{cm}^3$

$1\text{m}^3 = 1\,000\,000\text{cm}^3$

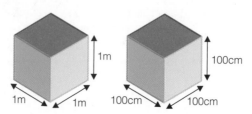

1. Work out the volumes of the following solids to 3 s.f. 🔲

   a)

   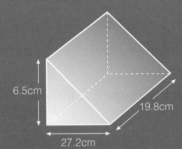

   6.5cm

   19.8cm

   27.2cm

   85cm

   b)

   10.6cm

2. Work out the surface area of a cuboid with height 6cm, width 4cm and length 10cm.

3. The volume of a cylinder is 2000cm³ and its radius is 5.6cm. Work out its height to 3 s.f. Use the π key on your calculator. 🔲

4. A prism has a volume of 10cm³. The prism is enlarged by a scale factor of 3. What is the volume of the enlarged prism?

   A 90cm³   **B** 27cm³   **C** 270cm³   **D** 900cm³   **E** 30cm³

5. The solid is a prism with height $5x$. Write an expression for the volume of the solid. Show your working and simplify your expression.

   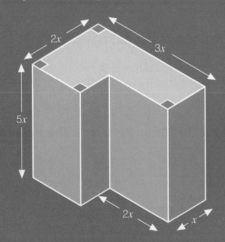

   $2x$

   $3x$

   $5x$

   $2x$

   x

   Not to scale

Progress Check

Worked questions

1. Mrs Hall's journey to work normally takes her 1 hour and 18 minutes. 📠

   a) Mrs Hall starts work at quarter to nine in the morning but it takes her 15 minutes to park her car and get to her desk.

   What is the latest time she can set off in the morning? Give your answer in the 24-hour clock. *(1 mark)*

   Time taken from home to desk = 1 hour 18 minutes + 15 minutes
   = 1 hour 33 minutes
   quarter to 9 − one hour = quarter to 8
   quarter to 8 − 30 minutes = quarter past 7
   quarter past 7 − 3 minutes = 12 minutes past 7 = 07.12
   8.45 − 1.33 = 7.12 = 07.12
   The latest time Mrs Hall can set off is 12 minutes past 7 (or 07.12).

   b) On a normal day, Mrs Hall's average speed is 45 kilometres per hour. What distance does she travel from home to work? *(2 marks)*

   Mrs Hall's journey = 45 × 1 hour 18 minutes
   Mrs Hall's journey = $45 \times 1\frac{18}{60}$

   $$= 45 \times 1\frac{3}{10} = 58.5km$$

   or 45 × 1.3 = 58.5km

   Mrs Hall travels 58.5km from home to work.

   c) Yesterday, because of a traffic jam, it took Mrs Hall 2 hours and 15 minutes to return home.

   What was the average speed of her journey home? *(2 marks)*

   Speed = 58.5 ÷ 2.25 = 26 kilometres per hour
   The average speed of Mrs Hall's journey home was 26 kilometres per hour.

2. A rugby pitch is laid out in a field 180m × 160m. The pitch itself is 105m × 76m.

   a) What area of the field remains for the spectators? 📠 *(3 marks)*

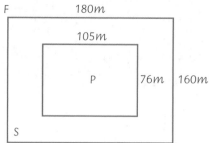

   Area of F = 180m × 160m = 28 800m²
   Area of P = 105m × 76m = 7980m²
   Area of S = 28 800 − 7980 = 20 820m²
   The area of the field remaining for the spectators is 20 820m².

b) As part of their warm-up, the players jog twice around the outside of the field and twice around the outside of the pitch.

How far do they jog altogether? Give your answer in km and m. 📱 *(3 marks)*

Perimeter of field = 2(180 + 160) = 680m
Perimeter of pitch = 2(105 + 76) = 362m
Distance jogged = (680 × 2) + (362 × 2) = 2084m = 2km 84m
The players jog 2km 84m altogether.

Take care with conversions, especially when zeros are involved and remember to make the conversion if the question asks.

Perimeter = 2 ($l + w$)

3. Joe's garden has a perimeter of 62m. The length of his garden is 18m. He wants to cover it with turf. He has a pond in one corner. Across the opposite corner is a triangular flower bed. 📱

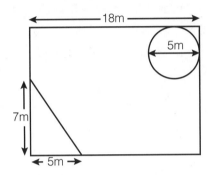

a) How many square metres of turf should he order? Round your answer to 1 d.p. *(4 marks)*

Area of T = area of whole – (area of P + Area of F)
Perimeter = 62 = 2(18 + w)
        = 31 = 18 + w
Width of garden = 31 – 18 = 13m
Area of garden = 18 × 13 = 234m²
P is a circle so, area of P = πr^2
        = π 2.5²
Area of P = 6.25 × π or 3.142 = 19.63495408 or 19.6375
        = 19.6m²

F is a triangle so, area of F = $\frac{1}{2}$ ($b \times h$)
Area of F = $\frac{1}{2}$ (7 × 5) = 17.5m²

Area of T = 234 – (19.6 + 17.5) = 196.9m²
Joe should order 196.9m² of turf.

Before you start, label the relevant areas. So P = pond, F = flower bed, T = turf.

Before you can find the area of the garden, you need to find the width.

Either use the π button on your calculator or multiply by 3.142

b) Joe also wants to tile the floor of his conservatory. The floor measures 8m by 6m. The tiles are 20cm × 15cm.

How many tiles does he need to buy? *(3 marks)*

Area of floor = 8 × 6 = 48m² (or 800 × 600 = 480 000cm²)
Area of one tile = 0.2 × 0.15 = 0.03m² (or 20 × 15 = 300cm²)
Number of tiles = 48 ÷ 0.03 = 1600 (or 480 000 ÷ 300 = 1600)
Joe needs 1600 tiles to fit the conservatory without overlap.

A bell should ring in your head because the units are different. You can either change the cm into m or vice versa. Try to avoid dealing with large numbers because you are more likely to make a mistake.

4. A block of wood measures 31.5cm × 19cm × 7cm. ▣

a) What is the surface area of the block of wood? *(4 marks)*

Area of whole = (area of A × 2) + (area of B × 2) + (area of C × 2)
Area of A = 31.5 × 19 = 598.5cm²
598.5 × 2 = 1197cm²
Area of B = 19 × 7 = 133cm²
133 × 2 = 266cm²
Area of C = 31.5 × 7 = 220.5cm²
220.5 × 2 = 441cm²
Area of whole = 1197cm² + 266cm² + 441cm² = 1904cm²
The surface area of the block of wood is 1904cm².

You could sketch the block of wood and label faces with the same dimensions with the same letter. For example, two of the faces could be labelled A, two could be labelled B and two could be labelled C.

b) Mary removes a cylinder of wood with a diameter of 4cm from the centre of the block.

What is the volume of the block with the cylinder removed? Round your answer to 1 d.p.

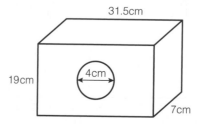

(3 marks)

Volume of new block = volume of original – volume of cylinder

Volume of original block = 31.5 × 19 × 7

= 4189.5cm³

Volume of cylinder = area of circle × length

= (π × 2²) × 7 = 3.142 × 4 × 7 = 87.976cm³

= 88.0cm³

Volume of new block = 4189.5 – 88

= 4101.5cm³

Volume of new block = 4101.5cm³

You have two volumes to calculate – the volume of the original block and the volume of the cylinder of wood.

c) The density of the cylinder of wood removed by Mary is 0.097g/cm³.

Calculate the weight of the cylinder to the nearest gram. *(2 marks)*

Density = $\frac{mass}{volume}$

Mass = density × volume

Mass of cylinder = 0.097 × 88

= 8.536

= 9 to the nearest gram

The cylinder of wood weighs 9 grams.

Practice questions

1. Here is part of a bus timetable for the 99 route from Newham to Lowermill.

Newham, Bus station	0640		10	40		1710
Newham, Train station	0647		17	47		1720
Tesco, Brook Street	0650		20	50		1725
Sycamore Ave	0653	then	23	53	mins past	1730
Stamford Road	0658	at	28	58		1737
Meadowlands	0703		33	03		1745
Waterhead	0705		35	05	each hour	1748
Tasker	0712		42	12		1757
Carrcote	0720		50	20		1809
Lowermill, The Square	0725		55	25	until	1816

   a) How long does the journey take from Meadowlands to Lowermill on the 0640 from Newham Bus Station? *(1 mark)*

   b) Waleed wants to arrive in Lowermill by 2pm.

   What is the latest bus he can catch from Sycamore Avenue? *(1 mark)*

   c) Jane arrives at the Waterhead stop at 1706 and just misses the 1705.

   How long will she have to wait for the next bus? *(1 mark)*

   d) How much longer is the 1710 journey from Newham Bus Station to Lowermill than the 1540 journey? *(1 mark)*

   e) The 99 bus travels at an average speed of 20km per hour on its morning journeys. How far does the bus travel from Newham Bus Station to Lowermill? *(2 marks)*

   f) What is the bus's average speed on its 1710 journey to Lowermill? Round your answer to the nearest kilometre per hour. *(2 marks)*

2. Three flags are being designed. 🖩

   a) The first flag will measure 1m × 1m and will be green, except for 1000cm², which will be red. What fraction of it will be green? *(1 mark)*

   b) The second flag will measure 38cm by 17cm. It will be white with a blue circle in the middle. The diameter of the circle will be 9cm.

   What will be the area of the white part? *(1 mark)*

   c) The third flag will be made up of three triangles as shown in the diagram below.

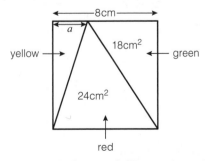

   The red triangle will have an area of 24cm². The green triangle will have an area of 18cm². What is the length of the base of the yellow triangle (marked *a*)? *(1 mark)*

3. Here is a plan of Terri's house and garden. 📱

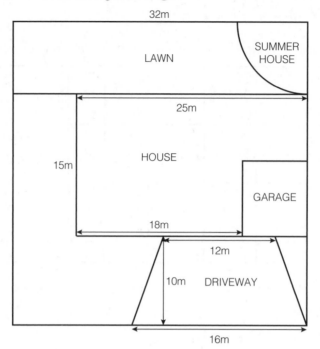

a) The area of the house is 312m². What are the dimensions of the garage? *(2 marks)*

b) The summerhouse sits on a quadrant with a radius of 5m.

What is the area of the lawn at the rear of the house? Round the area of the quadrant to 3 significant figures. *(2 marks)*

c) Calculate the area of the driveway. *(2 marks)*

d) The driveway is paved with flags each of which has an area of 900cm².

How many flags were used? Round your answer to the nearest whole one. *(1 mark)*

4. Complete the table by converting the given values into kilograms, grams or kilograms and grams. *(4 marks)*

kg	g	kg g
7.04kg		
	7004g	
		7kg 400g
0.07kg		

5. Write km², m², cm² or mm² after each of the following measures and then draw lines to match up those which are equal. One has been done for you. *(3 marks)*

1m² ⟶ 10000cm²

2   200

2   20000

2   2000000

6. Mia enlarged a photograph of her daughter by a scale factor of 2. The original photo measured 7cm by 4cm. What is the area of the enlargement? *(1 mark)*

7. This question is about three water tanks which are cuboid-shaped. 🖩

    a) There are 18m³ of water in a tank. If the tank is 2m long and 3m wide, what depth is the water? *(1 mark)*

    b) A tank half full of water has a layer of ice on top. The tank is 3m long and 0.5m wide.

    If the volume of the ice is 7500cm³, what is the thickness of the ice? *(1 mark)*

    c) 1m³ of water measures 1000 litres. A tank holds 200 litres of water when full. The tank's internal measurements are 250cm deep and 10cm wide.

    What is the length of the tank? *(1 mark)*

8. Chocolate raisins are packed into boxes measuring 70mm × 50mm × 20mm. The individual boxes are packed into a larger carton measuring 35cm × 25cm × 10cm.

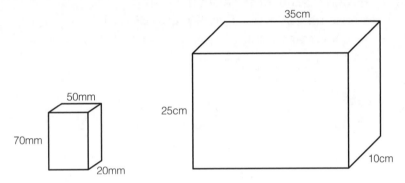

    a) How many boxes of raisins will each carton hold? *(3 marks)*

    b) Each larger carton is then wrapped in plastic shrink film. The roll of film is 550mm wide.

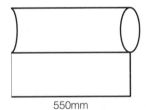

    What is the minimum length of film needed to cover each carton? *(2 marks)*

9. Here is a stainless steel bar shaped like a triangular prism. 🖩

    a) Calculate the volume of the bar. *(1 mark)*

    b) The bar has a mass of 901.6 grams. What is its density? *(2 marks)*

8 Probability

Learning Summary

After studying this chapter you should be able to:
- understand and use the probability scale from 0 to 1
- find the theoretical probability of an event
- calculate the probability of an event not happening
- find and record all mutually exclusive outcomes for single events and two successive events in a systematic way
- know when to add and multiply two probabilities
- understand and use set notation
- use and work out simple probabilities in Venn diagrams
- understand relative frequency as an estimate of probability.

8.1 The probability scale

Probability is the chance of something happening. All probabilities lie between 0 and 1 and can be written as fractions, decimals or percentages. Probabilities can be shown on a probability scale.

For example:

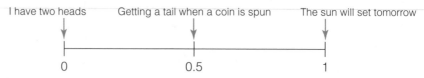

If you are answering questions on probability, always check that your answer is not greater than 1. If it is, it must be wrong.

Make up your own probability scale using statements about the weather. Think of a ridiculous statement which could never happen, some statements you are absolutely certain about and some in between and put them onto a scale. Think of the appropriate probability vocabulary to match. You could then try with another topic. Be as humorous as you can with the impossible ones!

Example

A bag contains 3 red, 6 green and 3 blue beads. If a bead is chosen at random:

a) mark with a P the probability of choosing a green bead

b) mark with a T the probability of choosing a red bead

c) mark with an X the probability of choosing a black bead.

X is at 0 since there are no black beads. A black bead will definitely not be chosen.

T is only a quarter of the way along the scale because only $\frac{1}{4}$ of the beads are red.

P is at 0.5. It has an evens chance of being chosen because half of the beads are green.

Exhaustive events account for all the possible **outcomes**. For example 1, 2, 3, 4, 5, 6 are all the possible outcomes when a fair dice is thrown.

Probability of an event happening

If we know what all the possible outcomes are, we can calculate the
theoretical probability of something happening:

Probability of an event = $\dfrac{\text{number of ways an event can happen}}{\text{total number of outcomes}}$

P(event) is the shortened way of writing the probability of an event.

Examples

1. The letters in the word MATHEMATICS are placed in a container, and
   a letter is taken out at random. What is the probability of taking out:

   a) a letter T?

   Since there are 11 letters, each of the probabilities are out of 11.

   $P(T) = \dfrac{2}{11}$

   b) a letter S?

   $P(S) = \dfrac{1}{11}$

   c) a letter R?

   Since there is no letter R, the probability is zero.

   $P(R) = 0$

2. Thomas has some coloured blocks, 3 red, 4 blue and 6 green, in a
   large bag. If he picks out a block at random, what is the probability
   that the block is:

   a) red?

   $P(red) = \dfrac{3}{13}$

   b) blue?

   $P(blue) = \dfrac{4}{13}$

   c) blue, green or red?

   $P(blue, green or red) = \dfrac{13}{13}$

   All the probabilities will add up to 1 here since Thomas will definitely choose a red, blue or green coloured block.

   d) white?

   $P(white) = 0$

Look in a dictionary for some long words which have two or three of the same letter and repeat the MATHEMATICS question opposite. As well as working out the probability of each letter being picked out, also work out the probabilities of each one not being drawn out. Remember P(event will happen) + P(event will not happen) = 1.

Probability of an event not happening

Mutually exclusive events cannot happen at the same time, for example getting a 6 and a 1 on one throw of a dice. For mutually exclusive events, the sum of the probabilities is 1:

P(event will not happen) = 1 − P(event will happen).

Examples

1. The probability that it will rain tomorrow is $\frac{2}{9}$

   What is the probability that it will not rain tomorrow?

   P(will not rain) = 1 − P(it will rain)

   P(will not rain) = 1 − $\frac{2}{9}$

   $\qquad\qquad\quad = \frac{7}{9}$

2. Some discs are placed in a bag. Most are marked with the number 1, 2, 3, 4 or 5. The rest are unmarked. The probability of picking out a disc marked with a particular number is:

   P(1) = 0.2

   P(2) = 0.1

   P(3) = 0.05

   P(4) = 0.15

   P(5) = 0.35

   What is the probability of picking a disc:

   a) marked with 2, 3 or 4?

   $\qquad$ P(2, 3 or 4) = (0.1 + 0.05 + 0.15)

   $\qquad\qquad\qquad = 0.3$

   b) not marked with a number?

   $\qquad$ P(not marked with a number) = 1 − P(marked with a number)

   $\qquad\qquad\qquad\qquad\qquad\qquad\quad = 1 − (0.2 + 0.1 + 0.05 + 0.15 + 0.35)$

   $\qquad\qquad\qquad\qquad\qquad\qquad\quad = 1 − 0.85$

   $\qquad\qquad\qquad\qquad\qquad\qquad\quad = 0.15$

You need to be able to subtract decimals so that you can answer these probability questions. See Topic 2.1.

1. Use one of these words or phrases to complete each statement.

   impossible   very unlikely   unlikely   evens   likely   certain

   a) It is _____ that 15 people in the same class have the same birthday.

   b) It is _____ that the Prime Minister will have two heads tomorrow.

   c) It is _____ that most people will wear a coat when it is raining.

2. A bag contains 6 red and 5 blue counters. If a counter is chosen at random from the bag, find the probability that the counter is:

   a) red

   b) blue

   c) red or blue

   d) yellow.

3. A pencil case contains 7 red and 4 black pens. A pen is taken out of the pencil case at random. What is the probability the pen is:

   a) red?

   b) blue?

   c) black?

4. A box of chocolates contains 15 hard centres and 13 soft centres. One chocolate is taken out of the bag at random. Work out the probability it will be:

   a) a hard centre

   b) a soft centre

   c) a biscuit.

Progress Check

5. The probability that a torch works is 0.63. What is the probability that the torch will not work?

6. 1000 tickets are sold in a raffle. Amy buys 12 tickets, Ruby buys 3 tickets, Colin buys 20 tickets and Rebecca buys 6 tickets. If there is only one winning ticket, what is the probability that:

   a) Ruby wins?

   b) Colin wins?

   c) A girl from this group wins?

   d) Amy will not win?

7. The probability that Sarah sends a text on any given day is 0.64. Work out the probability that Sarah does not send a text on any given day.

8. The probability that Shobna buys a chocolate bar is $\frac{7}{9}$. What is the probability that Shobna does not buy a chocolate bar?

8.2 Possible outcomes for two successive events

Lists, sample space diagrams, Venn diagrams and two-way tables are useful when answering probability questions with two successive events.

Lists

Making lists of possible outcomes of two events are useful but only when the items are written in an ordered way.

Examples

1. A coin can land in two ways: head up (H) or tail up (T). If the coin is tossed twice, make a list of the four possible ways that the coin can land in two throws. What is the probability of getting two heads?

   H   H
   T   T
   H   T
   T   H

   P(2 heads) = $\frac{1}{4}$

2. For her lunch Lin can choose a main course and a dessert from the options shown. List all the possible outcomes of her lunch. What is the probability that Lin will choose pizza and cake?

   | Pizza, yoghurt | Chicken pie, yoghurt | Fish, yoghurt |
   | Pizza, cake | Chicken pie, cake | Fish, cake |

   P(pizza and cake) = $\frac{1}{6}$

> Make a list of all the possible ways that a coin can land in three throws. Calculate the probabilities of one head occurring in the three throws; two heads occurring and three heads occurring.

Sample-space diagrams

Example

> A sample-space diagram can be used when two events happen at the same time. Use one to record all possible combinations when two dice are rolled simultaneously. Then complete another sample-space diagram showing the different totals that can be obtained from rolling the two dice. Work out the probabilities of each of the totals.

The hands on these two spinners are spun at the same time.

The two scores are added together.

a) Represent the outcomes on a sample-space diagram.

b) What is the probability of a score of 7?

   The probability of a score of 7 = $\frac{3}{16}$

c) What is the probability of a multiple of 3?

   The probability of a multiple of 3 = $\frac{5}{16}$

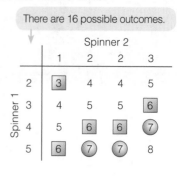

There are 16 possible outcomes.

		Spinner 2		
	1	2	2	3
2	3	4	4	5
3	4	5	5	6
4	5	6	6	7
5	6	7	7	8

Two-way tables

Example

The two-way table shows the number of students in a class who are left-handed or right-handed.

Hand	Male	Female	Total
Right	14	10	24
Left	2	7	9
Total	16	17	33

a) What is the probability that a person chosen at random is right-handed?

$$P(\text{right-handed}) = \frac{24}{33}$$

b) If a boy is chosen at random, what is the probability that he is left-handed?

$$P(\text{left-handed}) = \frac{2}{16} = \frac{1}{8}$$

You can sometimes simplify probabilities by cancelling down the fractions.
See Topic 1.2.

Make up your own two-way table. You just need to think of two variables, like gender and favourite school subjects (from a list of four). Make up some probability questions for a friend to answer. When you make up your own questions, you really begin to understand what probability is all about.

Set notation

A **set** is a collection of objects. The objects are called the elements or members of the set. Sets can be shown in a diagram called a **Venn diagram**.

Term	Meaning	Example
Set	A collection of items called members or elements	Set A = {1, 2, 3, ... 100} This set is all the whole numbers from 1 to 100.
Subset	A set made from members of a larger set	Set B = {2, 4, 6, ... 100} These are all the even numbers from 2 to 100.
Finite set	A given number of members of a set	Set C = {2, 4, 6, ... 20} These are the even numbers less than or equal to 20.
Infinite set	An unlimited number of members of a set	Set D = {2, 4, 6, ...} These are all the even numbers.
Empty set or **null set** $\varnothing$	A set that contains no members	
Universal set ϵ	This is a set that contains all possible members	

Union of sets (∪)

Set A ∪ B contains members belonging to A or B or both.

For example, if A = {2, 3, 4} and B = {4, 5, 6, 7}, then A ∪ B = {2, 3, 4, 5, 6, 7}. All members are in both sets.

Intersection of sets (∩)

Set A ∩ B contains members belonging to both A and B.

For example, if A = {2, 3, 4} and B = {1, 3, 5}, then A ∩ B = {3} since this is the only member in both sets.

If A and B have no members in common, then A ∩ B = ∅, i.e. it is an empty set.

Venn diagrams

Venn diagrams can be used to show information and to help solve probability questions.

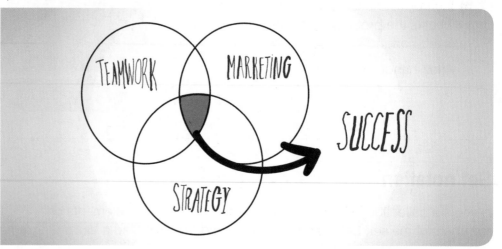

The universal set contains all the elements being discussed and is shown as a rectangle.

Relationships between sets can be shown on a Venn diagram.

Look at these examples:

A + A′ = ∈	A ∪ B	A ∩ B
![A' diagram: rectangle with 2 4 6 8 outside ellipse labelled A' containing 1 3 5, A, 7 9, and 10 in corner]	![A ∪ B diagram: overlapping circles A and B containing 2, 4, 3, 9, 6, 8, 12, 10, 15]	![A ∩ B diagram: overlapping circles A and B containing 3, 9, 6, 12, 15, 18, 2, 4, 8, 10]
For example: A = {1, 3, 5, 7, 9} A′ = {2, 4, 6, 8, 10} ∈ = {1, 2, 3, ...10}	For example: A = {3, 6, 9, 12, 15} B = {2, 4, 6, 8, 10, 12} A ∪ B = {2, 3, 4, 6, 8, 9, 10, 12, 15}	A = {3, 6, 9, 12, 15, 18} These are multiples of 3. B = {2, 4, 6, 8, 10, 12, 18} These are multiples of 2. A ∩ B = {6, 12, 18} These are multiples of 6

Example

Out of 40 students, 14 are taking French and 29 are taking German.

a) If five students are taking both subjects, how many students are taking neither subject?

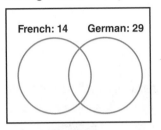

First draw the universal set for the 40 students, with two overlapping circles labelled with the total in each.

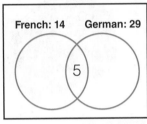

Since five students are taking both classes, put 5 in the overlap.

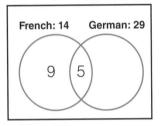

Five of the 14 French students have been accounted for, leaving nine students taking French but not German, so put 9 in the "French only" part of the "French" circle.

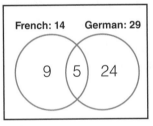

Five of the 29 German students have been accounted for, leaving 24 students taking German but not French, so put 24 in the "German only" part of the "German" circle.

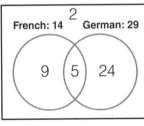

This tells us that a total of 9 + 5 + 24 = 38 students are taking either French or German (or both). This leaves two students unaccounted for, so they must be the ones taking neither subject.

So two students are taking neither subject.

b) How many students are taking at least one of the two subjects?

38 students ← Two students out of 40 are taking neither subject, so 38 are taking at least one.

c) What is the probability that a randomly-chosen student from this group is taking only German?

There is a $\frac{24}{40}$ = 0.6 = 60% probability that a randomly-chosen student in this group is taking only German.

You can write probabilities as fractions, decimals and percentages. You need to be able to convert between each one. See Topics 1.2 and 1.3.

The addition law

If two events (A and B) are mutually exclusive, the probability of A or B happening is found by adding the probabilities:

$$P(A \text{ or } B) = P(A) + P(B)$$

Example

There are 20 counters in a bag. 6 are red, 5 are white and the rest are blue. Find the probability that if Gill picks a counter at random, it is either red or white.

$P(\text{red}) = \frac{6}{20}$

$P(\text{white}) = \frac{5}{20}$

$P(\text{red or white}) = P(\text{red}) + P(\text{white})$

$\quad = \frac{6}{20} + \frac{5}{20}$ ← Red and white are mutually exclusive.

$\quad = \frac{11}{20}$

The multiplication law

Two events are said to be **independent** when the outcome of the second event is not affected by the outcome of the first event.

If two or more events are independent, the probability of A and B happening together is found by multiplying the separate probabilities:

$$P(A \text{ and } B) = P(A) \times P(B)$$

Example

The probability that it will be windy on any day in April is $\frac{3}{10}$. Find the probability that it will be:

a) windy on both April 1 and April 3.

$P(\text{windy and windy}) = \frac{3}{10} \times \frac{3}{10} = \frac{9}{100}$

b) windy on April 5 but not on April 20.

$P(\text{windy and not windy}) = \frac{3}{10} \times \frac{7}{10} = \frac{21}{100}$

You need to be able to multiply fractions and decimals. See Topic 1.2.

1. Reece has a pizza for his lunch. He has a choice of three toppings – mushroom, pineapple or ham. He chooses two toppings. Make a list of all the possible pizzas he can have.

2. Two fair dice are thrown at the same time and their scores are multiplied. Draw a sample-space diagram to show this information. Work out the probability of:

   a) a score of 3

   b) a score that is a multiple of 5.

3. The probability that Ashock does his homework is 0.8
   The probability that David does his homework is 0.45
   Find the probability that both boys do their homework. 🖩

4. The two-way table shows the number of infants who were immunised against an infectious disease and the number of infants who caught the disease.

	Immunised	Not immunised	Total
Did not catch the disease	83	8	91
Caught the disease	4	17	21
Total	87	25	112

   a) What is the probability that an infant who has been immunised catches the disease?

   b) What is the probability that an infant has been immunised?

5. The Venn diagram represents the subject choices of 60 students. Of these students, 6 students only chose Geography, 20 students chose both Geography and History and 16 students did not choose either Geography or History.

   a) Copy and complete the Venn diagram.

   b) Work out the value of x.

   c) How many students study Geography?

   d) What is the probability that a student chosen at random does not study Geography or History?

Progress Check

8.3 Estimating probability

You can estimate the probabilities of some events by doing an experiment. This is called estimated probability. The experiment must be repeated several times and a record kept of:

- the number of successful trials (when the event happens)
- the total number of trials.

$$\text{The estimated probability of an event} = \frac{\text{number of successful trials}}{\text{total number of outcomes}}$$

For example, the results of tossing a fair coin 100 times are recorded in the tally chart:

Top side of coin	Tally	Frequency
Head	卌 卌 卌 卌 卌 卌 卌 卌 卌 III	48
Tail	卌 卌 卌 卌 卌 卌 卌 卌 卌 卌 II	52

Using the results, the estimated probability of throwing a head

$$= \frac{\text{number of successful trials}}{\text{total number of trials}}$$

$$= \frac{\text{number of heads}}{\text{total number of tosses}}$$

$$= \frac{48}{100}$$

$$= 0.48$$

This is the estimated probability that the result will be a head.

The more times the experiment is carried out, the closer the estimated probability gets to the theoretical one.

The probability of some events can be predicted. For example, the probability of scoring 2 on a fair dice is $\frac{1}{6}$ because all the outcomes (1, 2, 3, 4, 5, 6) are equally likely. A predicted, or theoretical, probability can be used to estimate the expected number of successes in an experiment.

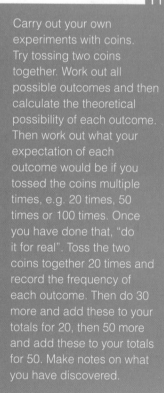

Carry out your own experiments with coins. Try tossing two coins together. Work out all possible outcomes and then calculate the theoretical possibility of each outcome. Then work out what your expectation of each outcome would be if you tossed the coins multiple times, e.g. 20 times, 50 times or 100 times. Once you have done that, "do it for real". Toss the two coins together 20 times and record the frequency of each outcome. Then do 30 more and add these to your totals for 20, then 50 more and add these to your totals for 50. Make notes on what you have discovered.

Examples

1. If a fair dice is thrown 300 times, approximately how many fives are likely to be obtained?

   $P(5) = \frac{1}{6} \times 300$

   $P(5) = 50$

   You would expect to get 50 fives.

2. The probability of obtaining a C grade in French at GCSE is 0.4
   If 200 students sit the exam, how many are expected to achieve a C grade?

   $P(C) = 0.4 \times 200$

   $= 80$

   Hence, 80 students are expected to achieve a C grade.

3. The probability that Alan is late to work is 0.2

   Alan works for 40 days before his next holiday. On how many days do you expect Alan to be late for work?

   $P(\text{late}) = 0.2 \times 40$

   $= 8$

   You would expect Alan to be late to work on 8 days.

Relative frequency

Some probabilities cannot be predicted, for example the probability that a piece of toast will land butter-side up. In this case we can repeat an experiment many times and find the **relative frequency** of the toast landing butter-side up.

If the toast lands butter-side up x times in n experiments, the relative frequency of the toast landing butter-side up is $\frac{x}{n}$

$$\text{Relative frequency of an event} = \frac{\text{number of successful trials}}{\text{total number of trials}}$$

The relative frequency of an event is used when you cannot calculate probabilities based on equally likely outcomes. Surveys allow you to estimate the results of a large group of people by finding the relative frequency with a smaller group.

For example, if a dice is thrown 180 times, it would be expected that about 30 twos would be thrown.

$$\frac{1}{6} \times 180 = 30$$

If we threw the dice 180 times and recorded the frequency of twos every 30 times, the results may look like the table below.

Number of throws	Total frequency of twos	Relative frequency
30	3	$\frac{3}{30} = 0.1$
60	7	$\frac{7}{60} = 0.12$
90	16	$\frac{16}{90} = 0.18$
120	19	$\frac{19}{120} = 0.16$
150	24	$\frac{24}{150} = 0.16$
180	31	$\frac{31}{180} = 0.17$

Make your own square-shaped spinner divided into quarters. Number the quarters 1, 2, 3 and 4. Start spinning the spinner, recording the number of times a 3 is obtained. Record the total number of spins and the total number of 3s obtained in a table (like the one opposite) every 10 spins, up to 100 spins. Then work out the relative frequency after each of those 10 spins and see how closely these match to the expected probability of obtaining a 3 (0.25).

Drawing a graph of the results shows that as the number of throws increases, the relative frequency gets closer to the expected probability.

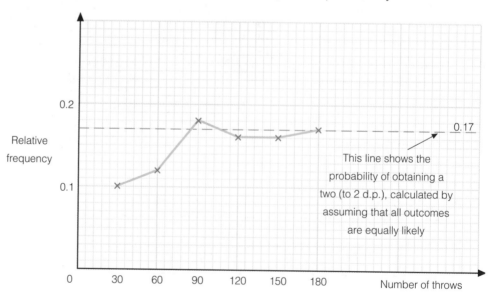

This line shows the probability of obtaining a two (to 2 d.p.), calculated by assuming that all outcomes are equally likely

1. The probability of passing a driving test at the first attempt is 0.65 If there are 200 people taking the test for the first time, how many do you expect to pass the test?

2. When a dice was thrown 320 times, a four came up 58 times. What is the relative frequency of getting a four?

3. The probability of getting flu this winter is $\frac{4}{9}$
   In a school of 1800 students, how many would you expect to get flu this winter?

Progress Check

4. Samuel thinks that his dice is biased. He throws the dice 120 times and these are his results.

Number on dice	1	2	3	4	5	6
Frequency	17	19	11	33	22	18

   a) Write down the number you think that the dice is biased towards? Explain your answer.

   b) What could Samuel do to make sure his results are more reliable?

Worked questions

1. Malik can choose one piece of fruit and one packet of crisps for a snack. His choices of fruit are apple, kiwi, banana or grapes. He may pick salty, cheese and onion or plain crisps.

   a) List all the possible outcomes for Malik's choice of snack.   *(2 marks)*

AS	KS	BS	GS
AC	KC	BC	GC
AP	KP	BP	GP

	Apple	Kiwi	Banana	Grapes
Salty	AS	KS	BS	GS
Cheese & Onion	AC	KC	BC	GC
Plain	AP	KP	BP	GP

   b) What is the probability of Malik choosing grapes?   *(1 mark)*

   There are 12 possible outcomes.

   Grapes occur in 3 of the outcomes.

   The probability of Malik choosing grapes, or P(G), is $\frac{3}{12}$

   $$\frac{3}{12} = \frac{1}{4}$$

   The probability of Malik choosing grapes is $\frac{1}{4}$.

   c) What is the probability of Malik choosing plain crisps?   *(1 mark)*

   $P(plain) = \frac{4}{12}$

   $= \frac{1}{3}$

   The probability of Malik choosing plain crisps is $\frac{1}{3}$.

2. A survey of favourite authors was carried out in Years 1, 2 and 3. The children could choose from Enid Blyton (EB), Roald Dahl (RD), Julia Donaldson (JD) and Michael Morpurgo (MM). The results were put into a two-way table:

	Year 1	Year 2	Year 3	Total
EB	6	0	4	
RD	4	4	6	
JD	7	6	2	
MM	4	2	5	
Total				

a) What is the probability of any random child choosing Julia Donaldson? *(1 mark)*

	Year 1	Year 2	Year 3	Total
EB	6	0	4	10
RD	4	4	6	14
JD	7	6	2	15
MM	4	2	5	11
Total	21	12	17	50

$P(JD) = \dfrac{15}{50}$

$\quad = \dfrac{3}{10}$

The probability of a random child choosing Julia Donaldson is $\frac{3}{10}$

With these types of tables you need to be careful about choosing which total you need to use. In this case, the question refers to all of the children. 15 chose Julia Donaldson out of a possible 50.

b) What is the probability of a random Year 1 child not choosing Julia Donaldson? *(1 mark)*

$P(\text{not choosing JD}) = 1 - P(JD)$

$P(\text{not choosing JD}) = 1 - \dfrac{7}{21}$

$\quad = 1 - \dfrac{1}{3}$

$\quad = \dfrac{2}{3}$

The probability of a random Year 1 child not choosing Julia Donaldson is $\frac{2}{3}$.

Here, you need to use the Year 1 total as your number of possible outcomes. Remember that the total of all probabilities equals 1.

c) Susie commented that Enid Blyton is just as popular in Year 1 as Julia Donaldson is in Year 2. Say whether she is right, giving a reason for your answer. *(1 mark)*

$P(EB - Y1) = \dfrac{6}{21} = \dfrac{2}{7}$

$P(JD - Y2) = \dfrac{6}{12} = \dfrac{1}{2}$

Susie is wrong because half the Year 2 children chose Julia Donaldson but only $\frac{2}{7}$ Year 1 children chose Enid Blyton.

You may be tempted to agree with this statement, since six Year 1 children chose Enid Blyton and six Year 2 children chose Julia Donaldson. However, when you look at the probabilities, you will see a very different picture.

d) What is the probability of a random child choosing Michael Morpurgo or Roald Dahl? *(1 mark)*

$P(MM) = \dfrac{11}{50}$

$P(RD) = \dfrac{14}{50}$

$P(MM \text{ or } RD) = \dfrac{11}{50} + \dfrac{14}{50}$

$\quad = \dfrac{25}{50} = \dfrac{1}{2}$

The probability of a random child choosing Roald Dahl or Michael Morpurgo is $\frac{1}{2}$.

Don't cancel at this point!

This is where you have to use the addition rule for events. As the question states a random child, the number of possible outcomes is 50.

Practice questions

1. In a bag of 15 sweets, 10 are lemon sherbets and 5 are spearmints.

    a) Picking a spearmint at random and picking a lemon sherbet at random are mutually exclusive events. Explain why this is so. *(1 mark)*

    b) Picking a spearmint at random and picking a lemon sherbet at random are also exhaustive events. Use the probability of picking a spearmint at random and the probability of picking a lemon sherbet at random to explain why. *(1 mark)*

2. Look at this diagram:

    SUPERSTARS   **OWLS**   **UNITED**   **LIGHTNING**   **THUNDER**   **MARVELS**

    a) There are six netball teams are in a mini league. Each team plays all the other teams. List all the games which will take place. *(2 marks)*

    b) The teams all also play in a knock-out tournament. Names are drawn at random from a hat. The teams were drawn in this order:

MARVELS v. OWLS

UNITED v. LIGHTNING

SUPERSTARS v. THUNDER

    Complete the table to show the probabilities of each team being drawn from the hat in this order. Choose the best probability description from those given below to complete the final column. The first one has been done for you. *(2 marks)*

    Even chance   Certain   Possible   Impossible

	Probability of name being drawn	Description
Marvels	$\frac{1}{6}$	Possible
Owls		
United		
Lightning		
Superstars		
Thunder		

    c) Using the list you made in part **a)**, what is the probability of Marvels and Owls being drawn to play the first game? *(1 mark)*

3. The table shows the probability of a dog-handler at an agility competition owning a border collie or another breed of dog. The dog-handlers are classed as either adults or children.

	Adults	Children
Border collie	0.6	0.2
Other	0.18	0.02

a) What is the probability that an owner chosen at random does not own a border collie? *(1 mark)*

b) There are 90 adult competitors with border collies. How many competitors are there altogether? *(1 mark)*

c) What is the probability of an owner chosen at random being a child? *(1 mark)*

4. Lily and Ellie are playing a game. They each have the same number of coins in their purses. They both take out a coin at random at the same time and add them up.

a) Complete the sample-space diagram to illustrate the totals of all the possible outcomes. *(3 marks)*

Lily's choice		1p	2p	5p	10p	20p
	20p					
	10p					
	5p					
	2p					
	1p					
				Ellie's choice		

b) What is the probability of the total being a two-digit number? *(1 mark)*

c) What is the probability of the total being a multiple of 10? *(1 mark)*

d) If the total of the two coins is an odd number, Ellie wins. If the total is an even number, Lily wins.

Complete this sentence:

Theoretically, _____ is likely to win overall. *(1 mark)*

5. Anil and Josh played another game. Anil picked a number from one bag at the same time as Josh picked a number from a second bag.

First bag   **Second bag**

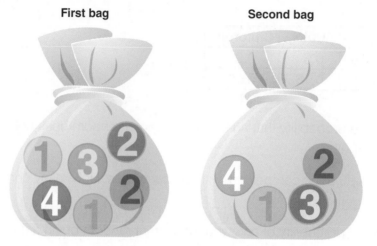

a) Complete the sample-space diagram to show all the possible outcomes.   *(3 marks)*

		1	2	3	4
4	4, 1			
3				
2				
2				
1				
1	1, 1			
		1	**2**	**3**	**4**

Anil's choice (left vertical label)

Josh's choice

b) What is the probability of Anil and Josh picking the same numbers?   *(1 mark)*

c) What are the least likely and most likely combination of numbers to be chosen?   *(1 mark)*

d) Design a bag containing six numbers (use only 1, 2, 3, 4) where there is a 50% chance of getting a 2.   *(1 mark)*

6. a) There are 35 coloured pencil crayons in Justin's pencil case. The probability that a crayon taken at random is red, black or blue is $\frac{2}{7}$.

   How many crayons are not red, blue or black?   *(1 mark)*

   b) Jack has 21 coloured pencil crayons in his pencil case. Five of them are orange, three are yellow and four are green. The rest are a variety of other colours.

   What is the probability that a crayon taken at random will be green or yellow?   *(1 mark)*

7. **a)** Show the following information in a Venn diagram.

Out of a class of 30 children, 15 are on the netball teams and 18 are on the basketball teams. Six children are on both teams. *(1 mark)*

b) How many children are not on either team? *(1 mark)*

c) What is the probability that a child chosen at random from the class will not be on either team? *(1 mark)*

8. In a quiz game each player spins a spinner to determine the category from which they will answer a question.

The table shows the number of times each category came up out of a total of 140 spins. ▣

	Geography	History	Science	Art	Sport
Frequency	29	18	30	35	28
Relative frequency					

Calculate the relative frequency of each category. Round your answers to 2 decimal places. *(2 marks)*

9. On four separate occasions, Lucy rolled a die and recorded her results below. ▣

	1st attempt	2nd attempt	3rd attempt	4th attempt
Number of rolls	20	40	60	120
Number of 2s	1	12	16	23
Relative frequency				

a) Calculate the relative frequency of getting a 2. Round your answers to 2 decimal places. *(2 marks)*

b) Which attempt is closest to the theoretical probability and why? *(1 mark)*

10. A bag contains red, blue, yellow and green tiddlywinks in the ratio of 1 : 2 : 3 : 6.

a) What is the probability of randomly pulling out a yellow tiddlywink? *(1 mark)*

b) Eight tiddlywinks are blue.

How many tiddlywinks does the bag contain altogether? *(1 mark)*

c) Sapna wants the probability of each colour being picked at random to be equal.

Which colour will have the same number in the new set up as in the old? *(1 mark)*

11. One fair dice is rolled twice.

a) What is the probability of rolling two sixes? *(1 mark)*

b) What is the probability of rolling two numbers which are not sixes? *(1 mark)*

c) What is the probability of rolling a six followed by not a six? *(1 mark)*

d) What is the probability of rolling a six and any other number in either order? *(2 marks)*

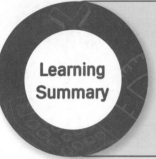

Learning Summary

After studying this chapter you should be able to:
- identify and obtain necessary information to solve a mathematical problem
- identify and select necessary data
- collect and record discrete data
- group data where appropriate into equal class intervals
- design a survey, questionnaire or experiment identifying possible sources of bias.

 9.1 Identifying and selecting data

Data

Every day people are bombarded with information called **data**. Data are often collected to test a **hypothesis**. A hypothesis is theory or explanation that has not been proved.

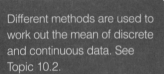

Different methods are used to work out the mean of discrete and continuous data. See Topic 10.2.

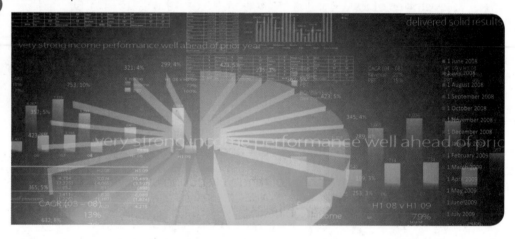

There are different types of data.

Discrete data	Can only take particular values. They are often found by counting. An example is the number of cars in a car park.
Continuous data	Can take any value in a given range. Such data are often found by measuring. Examples include the height and weight of year 8 students.
Primary data	Data that you collect yourself.
Secondary data	Data that somebody else has collected. For example, a census is carried out every 10 years in order to provide a 'snapshot' of people living in Britain. The census is a very rich source of data that is analysed to help the authorities plan for the future.

With an adult's permission, look up the definitions of the words 'discrete', 'secondary', 'primary' data on an Internet search engine. Check definitions and look at what kind of data are collected, how they are collected, and how they are organised and stored.

Bias

The word **population** is used to describe a set, collection or group of objects that are being studied. A sample is a small part of a population.

Anything that distorts the data in a sample, so that it will not give a representative picture of a population, is called **bias**. When collecting information it is important to make sure that there is no bias.

Bias usually occurs in one of these ways:

- The population or sample is not chosen correctly. An example of a biased sample would be whilst investigating homework trends at a school to only ask students in year 7 how much homework they do. This is biased because no other year groups have been taken into account.
- Through the style of questioning. For example, if your opinion is evident: "Most people want a new swimming pool. Do you want a new swimming pool?"

When identifying and selecting data, start with enough primary or secondary data so that a sample can be taken from it. Make sure that the sample is not biased. To avoid bias, ask a sample large enough to represent the whole population but small enough to be manageable.

For example, Rhysian is carrying out a survey into how often people do a sporting activity. She will need to ask sufficient people; 10 would be too small, so a sample of 50 people would be more reflective. She will need to ask a wide cross-section of people (i.e. a wide range of ages and both males and females).

Where Rhysian does her survey is also important. For example, her survey would be biased if she stood outside a leisure centre because the people she would ask are more likely to do a sporting activity.

1. Which of the following are primary data and which are secondary?
   a) Finding out information on a holiday destination by looking on the Internet.
   b) Measuring the height of all the students in your class.
   c) Finding out the shoe size of students in your class.
   d) Looking at records to see how many babies were born in January.
   e) Looking at tables of the number of road traffic accidents each year.

Progress Check

2. Explain why this sample is biased:
   'Investigating the pattern of absences for a school by studying the registers in February.'

9.2 Collecting data

There are some standard ways of collecting data.

Observations

An observation sheet (sometimes known as a data collection sheet) can be used to collect data. For example, here is an observation sheet used to test the hypothesis: 'Most members of staff at the school have a red car'.

When using an observation sheet, these points should be considered:
- Is the observation sheet clear and easy to use?
- Does the observation sheet actually answer the question asked?
- Was the data collected for long enough?
- Do the time and place of the observation affect the results?

Experiments

Experiments can be carried out in order to collect data. Important points to consider are:
- Does the experiment test the hypothesis?
- Have sufficient experiments been carried out to provide enough results to reflect what is happening?

For example, here is a biology experiment some students carried out to look at the growth of some seedlings.

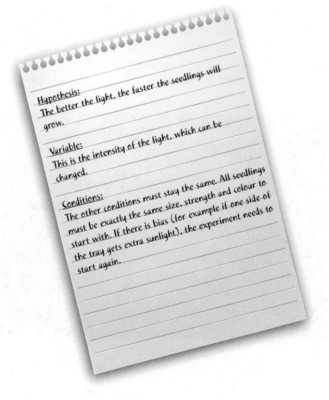

Hypothesis:
The better the light, the faster the seedlings will grow.

Variable:
This is the intensity of the light, which can be changed.

Conditions:
The other conditions must stay the same. All seedlings must be exactly the same size, strength and colour to start with. If there is bias (for example if one side of the tray gets extra sunlight), the experiment needs to start again.

Questionnaires

When designing or using **questionnaires**, the following points must be considered:

- Ask questions that cover the purpose of the survey.
- Keep the questions simple so that the answers are easy to analyse.
- Do not ask for information that is not needed, e.g. name or age.
- Make sure that your opinion is not evident, e.g. 'Do you agree that *Coronation Street* is better than *EastEnders*?'
- Give response boxes for all possible outcomes. They should not overlap.
- Include a time frame.

For example, look at the following question used in a questionnaire. It correctly includes a time frame (i.e. it specifies 'each week') and the values of the tick boxes do not overlap with each other (i.e. there is no confusion over which box to tick for a particular response).

> **How much do you spend on magazines each week?**
>
> Under £3 ☐
>
> £3 – £3.99 ☐
>
> £4 – £4.99 ☐
>
> £5 or more ☐

Design a short questionnaire that you could use to survey your classmates. Choose a survey topic that includes a time frame and ensure that the response boxes do not have overlapping values.

1. You are asked to do a survey outside a sports centre to find out about its popularity.

   a) One of the questions is: 'How old are you?'

   | Under 10 | 10–20 | 20–30 | 30–40 | Over 40 |
   | ☐ | ☐ | ☐ | ☐ | ☐ |

   Explain what is wrong with this question.

   b) Another question is: 'Do you go swimming?'

   | Sometimes | Occasionally | Often |
   | ☐ | ☐ | ☐ |

   Explain what is wrong with this question.

Progress Check

Tables and charts

Data that have been collected can be sorted by putting it into a table called a tally chart or a frequency table.

Tally charts

A tally chart shows the frequency of each item – in other words, how often the item occurs. A tally is a mark |. When the marks are grouped into fives they are easy to count. The fifth mark forms a gate ||||.

For example:

Hair colour	Tally	Frequency
Brown	‖‖ ‖‖	9
Ginger	‖‖	4
Black	‖‖ ‖	7
Blond	‖‖ ‖‖ ‖	12

Adding the tallies gives the frequency of each hair colour.

Frequency tables

If the data cover a large range of results, it is usual to group it into **class intervals**. Usually each class interval is the same width.

For example, in a test out of 50 the scores might be grouped as:

> The class intervals must not overlap.

0–10	11–20	21–30	31–40	41–50

It is sensible to choose class intervals of size 2, 5 or 10. For this example, the class intervals are not the same size because the first group (0–10) has a range of 11 and the others have a range of 10. A frequency table for the test might look like this:

Score	Tally	Frequency
0–10	‖‖	3
11–20	‖‖ ‖	7
21–30	‖‖	4
31–40	‖‖ ‖‖	10
41–50	‖	1

> Carry out a survey of the types of vehicles passing your house (e.g. car, lorry, bus, motorbike, etc.) in order to compile a tally chart of the information. Complete a frequency table using your tally chart.

Data such as the scores above are described as discrete because they can only take certain values. For example, you can have a score of 31 or 32, but not 31.4.

Continuous data can take any value within a certain range. For example, a person's height can take any value (within the range of heights that humans grow) – it doesn't have to be limited to a fixed height such as 1.5m or 1.6m because it can be measured to any value in between.

For continuous data, the class intervals are usually written using inequalities. For example, the table below shows the height in cm of 30 students.

$120 \leqslant h$ means the height can be equal to 120cm.

$120 \leqslant h < 130$ means that the heights are between 120 and 130cm.

Height (cm)	Tally	Frequency
$120 \leqslant h < 130$	𝍷𝍷𝍷𝍷 𝍷	6
$130 \leqslant h < 140$	𝍷𝍷𝍷𝍷	5
$140 \leqslant h < 150$	𝍷𝍷𝍷𝍷 𝍷𝍷𝍷𝍷 𝍷𝍷𝍷𝍷	14
$150 \leqslant h < 160$	𝍷𝍷𝍷𝍷	5

$h < 130$ means the height cannot be equal to 130cm. 130cm would be in the next class interval.

Inequalities are used to show continuous data. See Topic 3.2.

Two-way tables

Two-way tables are used to show two sets of information about the same group of individuals.

For example, a teacher has conducted a survey of the students in year 8 to find out their favourite subject:

Two-way tables are useful when finding the probability of two events. See Topic 8.2.

	Maths	English	Science	Total
Boys	20	10	15	45
Girls	30	20	10	60
Total	50	30	25	105

The table shows that 20 boys preferred Maths and 30 students preferred English.

Carry out a survey to collect data with two variables. For example, you could survey the types of vehicle passing your house again, but this time also observing if the driver was male or female. Organise your data into a two-way table.

1. A group of year 9 students in a mental arithmetic test scored these marks:

12	9	41	34	21	17	6	15	50	47
15	37	36	41	27	24	20	17	39	32
6	42	19	37	41	48	50	26	48	30

   a) Using class intervals 1–10, 11–20, 21–30, 31–40 and 41–50, construct a frequency table.
   b) Which class interval has the highest frequency?

2. A market researcher interviewed people travelling by car and by train.
   • 45 out of the 100 car travellers had travelled 20 miles or less.
   • Of the 250 people interviewed, 85 had travelled more than 20 miles.

   Use a two-way table to find out the number of people who had travelled more than 20 miles by train.

Progress Check

Worked questions

This is continuous data, i.e. it is found by measuring. For continuous data, the class intervals are usually written using inequalities. Class intervals for the data will look like this:

Large: $h \geqslant 430$mm

Standard: 430mm $> h \geqslant 390$mm

Medium: 390mm $> h \geqslant 350$mm

Small: 350mm $> h$

Having established the class intervals, construct the table. Organise the heights into the correct categories, using tally marks. Be organised with this and cross off each height as you enter it. Then count to ensure your tally marks equal 25. Finally complete the frequency column.

1. In agility competitions, dogs compete in classes according to their height:
   - Large – 430mm and over
   - Standard – less than 430mm but 390mm and over
   - Medium – less than 390mm but 350mm and over
   - Small – less than 350mm.

   Listed below are the heights in millimetres of 25 dogs. Using a frequency table, organise the heights into the correct categories. *(2 marks)*

390	465	461	350	391	398	443	328	346	449
472	335	401	429	347	349	431	396	458	430
432	412	385	351	438					

Height of dog (mm)	Tally	Frequency
$h \geqslant 430$	|||| ||||	10
$430 > h \geqslant 390$	|||| ||	7
$390 > h \geqslant 350$	|||	3
$350 > h$	||||	5

2. Jordan carried out a survey in Years 7 and 8 to find out which, if either, of the local football teams each student supported. He discovered that, of the 31 students who supported Newham United, 16 were in Year 7. Of the 59 students who comprised Year 7 and 8, 15 students supported neither team. Of 29 students in Year 8, only four supported Newham City.

   Put this information into a two-way table to find out how many students supported Newham City and how many from each year supported neither team. *(3 marks)*

This kind of information can be confusing at first but by carefully organising a table and inputting the values you have been given, finding the missing information becomes easier. The two variables are the years and who the students support.

Supports	Year 7	Year 8	Total
Newham United	16	15	31
Newham City	9	4	13
Neither	5	10	15
Total	30	29	59

The order in which the missing values can be worked out is:

13 i.e. $59 - (31 + 15)$

9, i.e. $13 - 4$

30, i.e. $59 - 29$

5, i.e. $30 - (16 + 9)$

10, i.e. $15 - 5$.

So 13 students support Newham City. 5 students from Year 7 and 10 students from Year 8 support neither team.

Practice questions

1. **a)** State whether each of the following data sources is primary or secondary.

    i) Pam went to the library to find out more information about her grandfather's war record. *(1 mark)*

    ii) Barbara conducted an experiment at home to see how long different types of chocolate took to melt. *(1 mark)*

    iii) Marcus carried out a survey amongst his family to find out which TV programmes they watched. *(1 mark)*

    iv) Jamilla went onto the internet to find out how many women held seats in parliaments across the world. *(1 mark)*

    b) Which of the above examples will capture continuous data? *(1 mark)*

2. Study the following questions taken from questionnaires. None would be considered to be a fair question. Attach a reason from the box to the letter of the question. *(2 marks)*

Too sensitive	Too wordy	Too ambiguous	Too biased

   a) What, in your opinion, are the strengths and weaknesses of the reading tests currently being used, as opposed to the proposed tests, in British primary schools?

   b) Have you ever copied other students' answers in an exam?

   c) How often do you jog?

   Very frequently   **Frequently**   **Regularly**   **Rarely**

   d) Most children who miss lessons will end up with poor jobs, won't they?

a)	
b)	
c)	
d)	

3. What is wrong with the following surveys?

   a) Jenny conducted a survey to see whether traffic lights might be needed at a particular junction. She conducted the survey from 8am to 9am on a Tuesday. This was her observation sheet: *(2 marks)*

Vehicles travelling towards town	Tally
Vehicle turned first left	
Vehicle turned first right	
Vehicle turned second left	
Vehicle turned second right	
Vehicle turned third left	
Vehicle turned third right	
Vehicle carried straight on	

Vehicles travelling away from town	Tally
Vehicle turned first right	
Vehicle turned first left	
Vehicle turned second right	
Vehicle turned second left	
Vehicle turned third right	
Vehicle turned third left	
Vehicle carried straight on	

b) Joan wanted to find out how many adults might be non-swimmers. She wanted to separate the information into men and women. *(2 marks)*

She stood outside the swimming baths and asked people entering the building whether they could swim. This was her observation sheet:

Swimmers	Non-swimmers

c) Sally wanted to see how wearing a rug in winter affected horses' weights. The horses wearing rugs were on a hillside 380m above sea level. The horses without rugs were in a field with a field shelter. *(1 mark)*

4. Here are the times (*t*) in seconds scored by competitors in a dog agility competition:

24.17	30.69	35.72	35.83	24.01	29.73	34.09	26.04	34.6	31.4
31.14	25.32	27.5	33.45	30.8	29.3	24	34.1	28.61	27
30	32.64	27.52	30.1	28.4	32.99	33	32.54	31.9	32.68

a) Using class intervals of $24 \leqslant t < 27$, $27 \leqslant t < 30$, $30 \leqslant t < 33$ and $33 \leqslant t < 36$, construct a frequency table. *(3 marks)*

Time (*s*)	Tally	Frequency

b) How many competitors were in the class interval representing the fastest times? *(1 mark)*

c) What fraction is this of the total number of competitors? *(1 mark)*

5. Millie conducted a survey at a bird reserve. She wanted to know how many of the visitors were members of the Royal Society for the Protection of Birds (RSPB) and how many had visited the site before.

The rain damaged her observation sheet. She could remember that 45 of the 210 who were visiting the site for the first time were members of the RSPB. She could also remember that out of a total of 570 visitors in all, 54 non-members had visited the site before.

a) Complete this two-way table to help Millie to retrieve her missing data. *(3 marks)*

	RSPB member	Non-member	Total
1st visit	45		210
Visited before		54	
Total			570

b) What percentage of people who had visited before were RSPB members? *(1 mark)*

c) What percentage of the total number of visitors were non-members?

Round your answer to the nearest whole number. ◉ *(1 mark)*

After studying this chapter you should be able to:
- select and represent data using a variety of statistical diagrams
- identify which statistical diagram is most useful for a problem
- interpret statistical diagrams
- find and interpret averages of different data to compare distributions
- estimate and find the median and interquartile range for large data sets
- draw and interpret cumulative frequency diagrams.

Learning Summary

10.1 Statistical diagrams

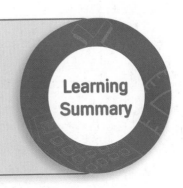

Data can be shown in several different types of diagram.

Pictograms

Pictograms use symbols: each symbol represents a certain number of items.

When drawing pictograms, make sure that:
- each row is labelled
- each symbol is the same size, with equal gaps between them
- a key is given.

Carry out a survey of the birds which visit your garden during a particular time span and use a pictogram to show your results. Explain it to another member of your family.

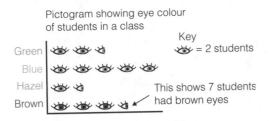

Pictogram showing eye colour of students in a class

Key
👁 = 2 students

This shows 7 students had brown eyes

Bar charts

A **bar chart** is a set of bars of columns of equal width. Bar charts are drawn with gaps between the bars. The height of each bar shows the frequency. Bar charts are used to show discrete data. Frequency is always on the vertical axis. A compound bar chart can also be used to compare two or more sets of data.

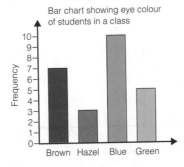

Bar chart showing eye colour of students in a class

Example

Thomas has carried out a survey of some students' favourite sport. Here are his results.

Favourite sport	Number of boys	Number of girls
Swimming	6	8
Football	15	3
Hockey	5	10
Tennis	9	14

a) Draw a dual bar chart of these results.

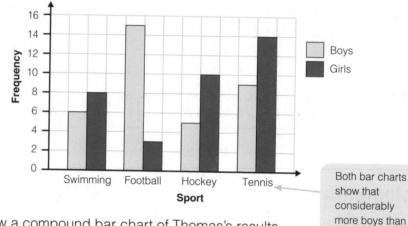

b) Draw a compound bar chart of Thomas's results.

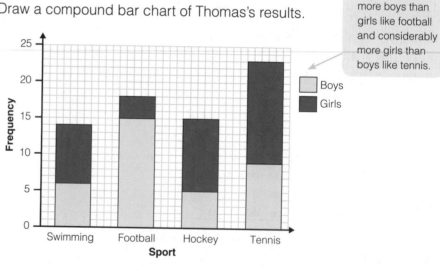

Both bar charts show that considerably more boys than girls like football and considerably more girls than boys like tennis.

Construct a bar chart of your own. You could do this easily by carrying out a survey of books in your bedroom. For example, you could set four categories such as 'fiction', 'educational', 'sport' and 'other'. You could then make the task a little more difficult by including another variable (such as 'hardback' or 'softback') so that you can create a dual bar chart or a compound one.

Sometimes lines are used instead of bars. These graphs are known as bar line graphs.

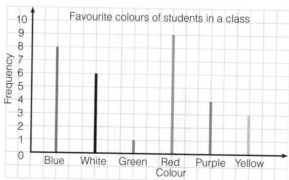

If the data is continuous, the bars must touch. This is known as a frequency diagram or histogram. The data must be grouped into equal class intervals if the length of the bar is used to represent frequency.

For example, the masses of 30 workers in a factory are shown in the table.

Mass (M kg)	Frequency
$45 < M \leq 55$	7
$55 < M \leq 65$	13
$65 < M \leq 75$	6
$75 < M \leq 85$	4
Total	**30**

Remember the following:

- The axes do not need to start at zero. Attention is usually drawn to this fact by using a jagged line like this.

- The axes are labelled and the graph has a title.

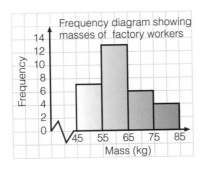

The diagram shows that the largest group of workers has a mass between 55 and 65kg.

Pie charts

In a pie chart the data is shown in a circle, which is split up into sections. Each section represents a certain number of items.

Drawing pie charts

When calculating the angles for a pie chart:

- find the total of the items listed
- find the fraction of the total for each item
- multiply the fraction by 360° to find the angle.

For example, the table gives the hair colour of 24 ten-year-olds.

Hair colour	Frequency
Brown	8
Auburn	4
Blond	6
Black	6
Total	**24**

8 out of 24 have brown hair so $\frac{8}{24} \times 360° = 120°$

Key in on the calculator:

Auburn: $\frac{4}{24} \times 360° = 60°$

Blond: $\frac{6}{24} \times 360° = 90°$

Black: $\frac{6}{24} \times 360° = 90°$

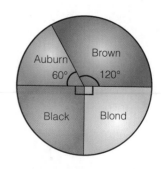

> Create a pie chart using the information that you obtained when creating a bar chart in the previous activity. Be as accurate as you possibly can when measuring the angles.

> You can set up fractions of a quantity to work out angles in a pie chart. See Topic 1.2.

Interpreting pie charts

Sometimes you will be given a pie chart and asked to work out the numbers it represents.

Example

The pie chart shows how some students spent Saturday night. If 140 students went to the ice rink, how many went to the disco and the cinema?

80° represents 140 students
1° represents $\frac{140}{80} = 1.75$ students
Number at disco = 160° × 1.75 = 280 students
Number at cinema = 120° × 1.75 = 210 students

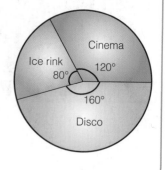

> You need to be able to multiply with decimals for questions like these. See Topic 2.1.

Comparing pie charts

It is important that pie charts can be accurately compared. When comparing pie charts, you need to be able to compare the proportions and not just the size of the angle.

Example

A census of two towns was carried out to look at the proportions of age within each town.

The pie charts show the results.

> Notice that each pie chart represents a different number of people, so we cannot just compare the angles.

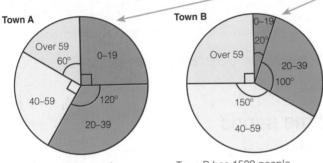

Town A has 3750 people.   Town B has 1500 people.

Ahmed says: 'There are more people aged over 59 living in Town B because the angle of the pie chart is bigger'. Is Ahmed correct? Give a reason for your answer. ▦

In Town A, $\frac{60°}{360°}$ represents the fraction of people aged over 59.

$\frac{60}{360} \times 3750 = 625$ people aged over 59 in Town A.

In Town B, $\frac{90}{360} \times 1500 = 375$ people aged over 59.

Ahmed is wrong – there are more people aged over 59 in Town A than in Town B. Town A has a greater population than Town B, meaning that the 60° angle in the first pie chart represents a greater number of people than the 90° angle in the second pie chart.

Frequency polygons

To draw a **frequency polygon**, join the midpoints at the top of each bar in the frequency diagram.

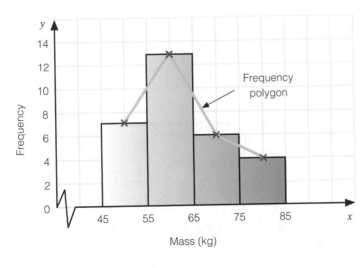

Mass (kg)

Frequency polygons can be superimposed on top of each other to compare results.

For example, these two frequency polygons show the distances jumped in a high-jump competition by students in year 7 and year 9.

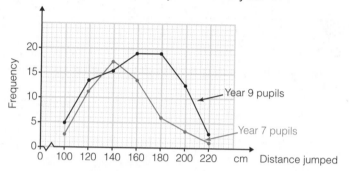

From the polygons we can see that in general a greater proportion of students from year 9 jumped greater heights.

Line graphs and time series

You need to be able to read graph scales accurately and plot the points with a small cross. See Topic 4.2.

Line graphs are a set of points joined by lines. Line graphs can be used to show continuous data, and show how a quantity changes over time.

For example:

Year	2007	2008	2009	2010	2011	2012	2013	2014
Number of cars sold	420	530	480	560	590	620	490	440

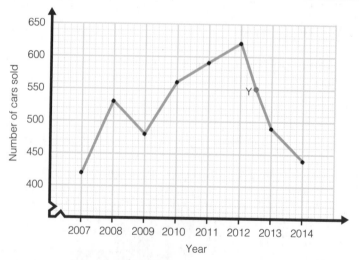

The middle values (for example at Y) have no meaning. Point Y does not mean that halfway between 2012 and 2013, 550 cars were sold.

A **time series** is made up of numerical data recorded at intervals of time and plotted as a line graph.

This diagram shows a time series showing seasonal fluctuations.

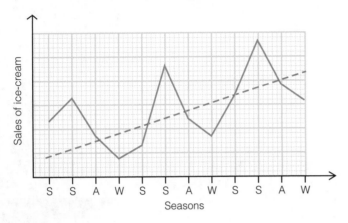

Scatter graphs

A **scatter graph** (also known as a scatter diagram or a scatter plot) shows the correlation between two sets of data.

Correlation

Correlation is a measurement of how strong the relationship is between two sets of data. There are three types of correlation:

Positive correlation	Negative correlation	Zero or no correlation
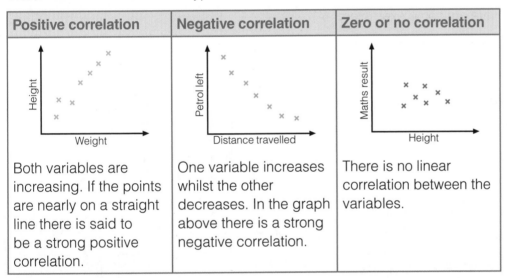		
Both variables are increasing. If the points are nearly on a straight line there is said to be a strong positive correlation.	One variable increases whilst the other decreases. In the graph above there is a strong negative correlation.	There is no linear correlation between the variables.

The line of best fit

The line of best fit is the line that best shows the trend of the data. The line goes in the direction of the data and has roughly the same number of points above it as below it. A line of best fit can be used to make predictions.

For example, the table below shows the Maths and History results of 11 students.

Maths test (%)	64	79	38	42	49	75	83	82	66	61	54
History test (%)	70	36	84	70	74	42	29	33	50	56	64

The data is plotted on a scatter graph, which suggests that there is a strong negative correlation – in general, the better the students did in Maths the worse they did in History, and vice versa.

You can work out the equation of the line of best fit by using the equation $y = mx + c$, where m is the gradient and c is the constant. See Topic 4.2.

The line of best fit can be used to predict Amy's Maths result if she scored 78% in History. Amy's estimated Maths result is approximately 43%.

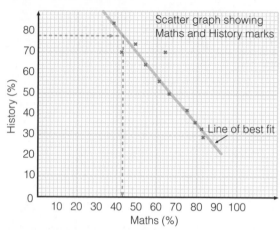

Scatter graph showing Maths and History marks

Misleading graphs

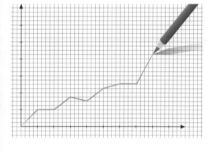

Statistical graphs are sometimes misleading – they do not always tell the true story. Misleading graphs are often seen when sales of products are being advertised.

Here are some examples:

This graph is misleading because it has no scales and the bars are not the same width.	
This graph is misleading because the scales do not start at zero. As a result, the differences between the bars look much bigger than they actually are.	
This pictogram is misleading because the pictures change size. Although Brand B has only sold twice the amount of Brand A, it gives the impression of having sold much more.	

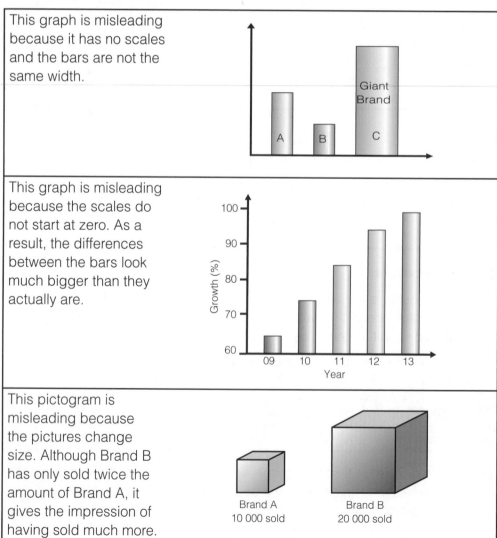

1. The pie chart shows the favourite subjects of 720 girls.

   a) How many girls like Maths?

   b) How many girls like Art?

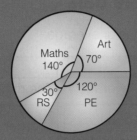

2. **a)** What type of correlation does the scatter graph show?

   b) Draw on the line of best fit.

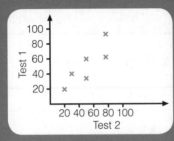

3. Explain why this graph is misleading.

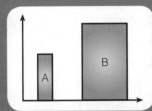

Progress Check

10.2 Averages and range

Averages of discrete data

You should know three types of average:

Mean	Median	Mode
Sometimes known as the 'average' $\text{Mean} = \dfrac{\text{the sum of a set of values}}{\text{the number of values used}}$	The middle value when the data is put in order of size If there are two numbers in the middle, the median is halfway between them	The value that occurs the most often

The **range** tells you the spread of the data.

Range = highest value – the lowest value

Examples

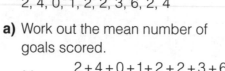

1. A football team scored the
   following number of goals in
   their first ten matches:
   2, 4, 0, 1, 2, 2, 3, 6, 2, 4

   a) Work out the mean number of
   goals scored.

   $$\text{Mean} = \frac{2+4+0+1+2+2+3+6+2+4}{10}$$

   $$= \frac{26}{10}$$

   $$= 2.6 \text{ goals}$$

   b) Work out the median number of goals scored.

   0, 1, 2, 2, 2, 2, 3, 4, 4, 6 ← Put in order of size first.

   Median = 0̸, 1̸, 2̸, 2̸, 2, 2, 3̸, 4̸, 4̸, 6̸

   $$= \frac{2+2}{2}$$

   $$= 2 \text{ goals}$$

   c) Work out the mode.

   Mode = 2 goals ← 2 was the most common number of goals to be scored.

   d) Work out the range.

   Range = 6 − 0 = 6

2. The mean of four numbers is 20. The mean of six other numbers is 36. What is the mean of all ten numbers?

The sum of the four numbers is $4 \times 20 = 80$ $\left(\frac{80}{4} = 20\right)$

The sum of the six numbers is $6 \times 36 = 216$ $\left(\frac{216}{6} = 36\right)$

Mean of all ten numbers is $\frac{80 + 216}{10}$

$= \frac{296}{10}$

$= 29.6$

The mean is useful when a typical value is wanted. It should not really be used if there are **outliers** – extreme values in a set of data. For example, 65 would be an outlier in the data 1, 2, 3, 4, 65 because it is significantly different from the rest of the values.

The **median** is a useful average when there are extreme values.

The mode is useful when the most common value is needed.

Finding averages from a frequency table

A frequency table tells you how many items are in a group.

Example

The table shows the number of sisters of students in a Year 7 class. 🖩

Number of sisters (x)	0	1	2	3	4	5
Frequency	4	9	3	5	2	0

a) Find the mean.

Mean $= \dfrac{\text{total of the result of frequency} \times \text{number of sisters}}{\text{total of the frequency}}$

$= \dfrac{(4 \times 0) + (9 \times 1) + (3 \times 2) + (5 \times 3) + (2 \times 4) + (0 \times 5)}{4 + 9 + 3 + 5 + 2 + 0}$

$= \dfrac{38}{23} = 1.65$ (2 d.p.)

b) Find the median.

There are 23 people altogether; the middle person is the 12th one. Looking at the table, the 12th person has one sister. So the median = 1.

c) Find the mode.

This is the number of sisters with the highest frequency, that is one sister.

d) Find the range.

$4 - 0 = 4$ sisters

Create three sets of cards, each numbered with the digits from 0 to 9. Shuffle them and turn four cards over to show the digits. Work out the mode, median, mean and range of the set of four. Then repeat the exercise with six or eight cards. You could even make up some puzzles for a friend. Choose eight cards but turn two of them face down for your friend, leaving the other six with the digits showing. Tell your friend what the range, mode, median or mean is. They have to work out what the two missing numbers are.

Make your own frequency tables using words in a reading book or magazine article. Count any 100 words and then count how many letters are in each of those 100 words. Construct a frequency table. From your frequency table, work out the mean number of letters per words. You could then find the range. Try comparing one set of words with another.

Stem-and-leaf diagrams

Stem-and-leaf diagrams are another way of recording information and they can be used to find the mode, median and range of a set of data.

> ### Example
>
> Here are some marks gained by some students in a Maths examination:
>
> 24 61 55 36 42 32 60 51 38 58 55 52 47 55 55
>
> **a)** Put the information into a stem-and-leaf diagram.
>
>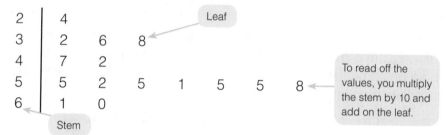
>
> Key: 2 | 4 means 24
>
> Rewriting in order gives:
>
> ```
> 2 | 4
> 3 | 2  6  8
> 4 | 2  7
> 5 | 1  2  5  5  5  5  8
> 6 | 0  1
> ```
>
> Key: 2 | 4 means 24
>
> **b)** Use the stem-and-leaf diagram to find the mode, median and range.
>
> Mode = 55
>
> Median is at the eighth score, i.e. 52.
>
> Range = 61 – 24 = 37

Averages of grouped data

When data are grouped, the exact values are not known. An estimate of the mean can be calculated by using the midpoint of the class interval. The midpoint is the halfway value.

For example, the heights of some Year 9 students are shown in the table below.

Height (h cm)	Frequency (f)	Midpoint (x)	$f \times x$
$140 \leqslant h < 145$	4	142.5	570
$145 \leqslant h < 150$	7	147.5	1032.5
$150 \leqslant h < 155$	14	152.5	2135
$155 \leqslant h < 160$	5	157.5	787.5
$160 \leqslant h < 165$	2	162.5	325

$$\text{Mean} = \frac{\sum fx}{\sum f}$$

> Σ is the Greek letter sigma and means 'sum of'.

$$\text{Mean} = \frac{(142.5 \times 4) + (147.5 \times 7) + (152.5 \times 14) + (157.5 \times 5) + (162.5 \times 2)}{4 + 7 + 14 + 5 + 2}$$

$$= \frac{4850}{32}$$

$$= 151.6\text{cm (1 d.p.)}$$

Because the data are grouped, the modal class is used instead of the mode.

Here the modal class is $150 \leqslant h < 155$ as it is the class interval with the highest frequency.

For grouped data, we can find the class interval containing the median.

There are 32 people in the survey; the middle person is between the 16th and 17th persons. Both have heights in the class interval $150 \leqslant h < 155$. So the class interval containing the median is $150 \leqslant h < 155$.

Finding the mean from a frequency diagram

You might be asked to estimate the mean from a frequency diagram.

Example

The frequency diagram shows the heights of some students. Estimate the mean height.

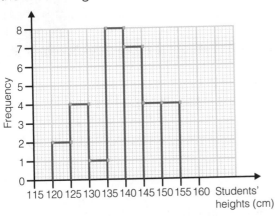

> First work out the midpoints and frequency of each bar. Remember to use the midpoint multiplied by the frequency.

$$\sum fx = (122.5 \times 2) + (127.5 \times 4) + (132.5 \times 1) + (137.5 \times 8) + (142.5 \times 7) + (147.5 \times 4) + (152.5 \times 4)$$

$$\frac{\sum fx}{\sum f} = \frac{4185}{30}$$

Mean height = 139.5cm

> Find out the heights of a group of at least 12 friends. Put your results into a frequency table like the one on page 198 and construct a frequency diagram. Estimate the mean height for your group of friends.

Comparing sets of data

The range and averages are used to compare sets of data.

For example, the students in 8M obtained a mean of 57% in a test. The top mark was 100% and the bottom mark 21%.

The students in 8T obtained a mean of 84% in the test. The top mark was 94% and the bottom mark 76%.

From the mean average, 8T performed better than 8M. Now look at the range for each class:

8M = 100% − 21% = 79%

8T = 94% − 76% = 18%

The range shows that 8M's marks were much more widely spread than 8T's. Some students in 8M obtained higher marks than those in 8T, even though the mean for 8T was higher.

You need to be able accurately work with numbers, fractions and decimals. See Topics 1.1 and 1.2.

Progress Check

1. The heights in centimetres of some students are:
   154, 172, 160, 164, 168, 177, 181, 140, 142, 153, 154, 153, 162
   a) Draw a stem-and-leaf diagram for this information.
   b) What is the range?
   c) What is the median?

2. The length of the roots of some plants is recorded in the table.
   a) Find an estimate for the mean length.
   b) What is the modal class?

Length (l cm)	Frequency	Midpoint (x)
$0 \leqslant l < 5$	6	2.5
$5 \leqslant l < 10$	9	
$10 \leqslant l < 15$	15	
$15 \leqslant l < 20$	9	
$20 \leqslant l < 25$	6	
$25 \leqslant l < 30$	2	

29 **10.3** Cumulative frequency graphs

Cumulative frequency graphs are very useful for finding the median and the spread of grouped data. Before drawing the graph, the **cumulative frequencies** have to be obtained by adding together the frequencies to give a running total.

For example, the table shows the time in minutes for 49 students' journeys to school.

Time (t minutes)	Frequency	Time (t minutes)	Cumulative frequency
$0 \leqslant t < 10$	15	$0 \leqslant t < 10$	15
$10 \leqslant t < 20$	16	$0 \leqslant t < 20$	31 (15 + 16)
$20 \leqslant t < 30$	9	$0 \leqslant t < 30$	40 (31 + 9)
$30 \leqslant t < 40$	6	$0 \leqslant t < 40$	46 (40 + 6)
$40 \leqslant t < 50$	3	$0 \leqslant t < 50$	49 (46 + 3)

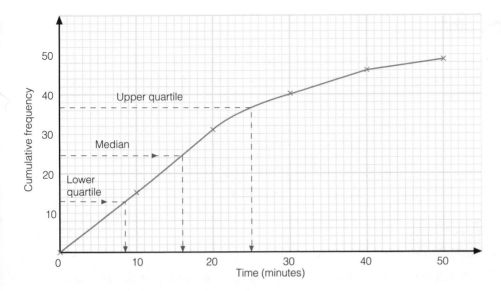

When drawing the graph, follow these steps:
*   When plotting the points, the upper class limits must be plotted. Plot (10, 15), (20, 31), (30, 40), etc.
*   Since no-one took less than zero time, the graph starts at (0, 0).
*   Join the points with a smooth curve.

Finding the median

The cumulative frequency curve can be used to estimate the median. The **median** is the middle value of the distribution. For the journey time data above:

Median = $\frac{1}{2}$ × total cumulative frequency = $\frac{1}{2}$ × 49 = 24.5

Find 24.5 on the vertical scale and read across to the curve and then down. This shows the median = 16 minutes.

Finding and using the interquartile range

You need to know the lower quartile, upper quartile and interquartile range of a set of data.

Upper quartile	Lower quartile
This is the value three-quarters of the way into the distribution.	This is the value one-quarter of the way into the distribution.
For the journey time data above that is $\frac{3}{4}$ × 49 = 36.75th value. By reading across at the appropriate place on the vertical scale, the upper quartile for this data = 25 minutes	For the journey time data above that is $\frac{1}{4}$ × 49 = 12.25th value. By reading across at the appropriate place on the vertical scale, the lower quartile for this data = 8.5 minutes

The interquartile range = upper quartile − lower quartile

So the interquartile range for the journey times = 25 − 8.5 = 16.5 minutes.

A large interquartile range indicates that the data is widely spread. A small interquartile range indicates that the data is concentrated about the median.

Find the lower quartile, upper quartile and interquartile range for the number of goals scored per game in the four professional English football divisions (i.e. approximately 46 matches) over a particular weekend.

Box plots

A box plot shows the interquartile range as a box, which makes it useful when comparing distributions. This box plot shows the journey time data from page 201.

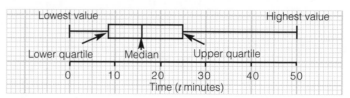

Example

The times in seconds taken by 11 students to solve a puzzle are listed in order:

2   4   4   5   5   7   8   8   9   11   12

Draw a box plot of this data.

2, 4, 4, 5, 5, 7, 8, 8, 9, 11, 12

Lower quartile   Median   Upper quartile

Time (t seconds)

Josie carried out a survey for her geography coursework. She recorded the distance travelled to an out-of-town shopping centre. Her results are shown in the table.

Progress Check

Distance (d miles)	Frequency
$0 \leqslant d < 5$	15
$5 \leqslant d < 10$	60
$10 \leqslant d < 15$	67
$15 \leqslant d < 20$	30
$20 \leqslant d < 25$	22
$25 \leqslant d < 30$	6

1. Draw a cumulative frequency graph.
2. Work out for this data:
   a) the median
   b) the interquartile range.

Worked questions

1. This table shows how many children there are in each household in Padcombe Green. 🖩

Number of children	0	1	2	3	4	5	6
Frequency	10	6	9	6	3	1	0

a) Find the mean. *(2 marks)*

Number of children	0	1	2	3	4	5	6	Total
Frequency	10	6	9	6	3	1	0	35 households
Total (number of children × frequency)	0	6	18	18	12	5	0	59 children

Mean = average number of children per household
$$= 59 \div 35$$
$$= 1.685$$
Mean = 1.69 (2 d.p.)

This type of question may or may not ask you to find a total. Even if it doesn't, this should be the first thing you should do with a frequency table. Frequency in this question is the number of households. Sometimes it helps to say to yourself, "10 households have no children, 6 households have 1 child, etc."

b) Find the mode. *(1 mark)*

The mode is 0.

The mode is the class with most entries.

c) Find the median. *(1 mark)*

The 18th number is in the 2 class, so the median is 2.

The frequency total is 35 so the middle value is the 18th. Move along the classes adding up until you reach the 18th number.

d) Find the range. *(1 mark)*

The range is 5.

The range goes from 0 children to 5 children, there being no households with 6 children.

2. This table shows the marks of 32 students in a science test. 🖩

Marks	Number of students (frequency)
1–5	3
6–10	6
11–15	12
16–20	9
21–25	2

a) Calculate the mean score for this group of students. *(2 marks)*

Marks	Number of students (frequency)	Midpoint	Midpoint × Frequency
1–5	3	3	9
6–10	6	8	48
11–15	12	13	156
16–20	9	18	162
21–25	2	23	46
Total	**32**		**421**

Because the scores are given within a range, you need to find the midpoint of the range of marks in each class interval and multiply it by the frequency. Then you add the midpoint frequencies and divide that total by the number of students.

$421 \div 32 = 13.16$ (2 d.p.)

The mean score for this group of students is 13.16.

b) Complete a cumulative frequency column for this data.   *(1 mark)*

Marks	Number of students (frequency)	Cumulative frequency
1–5	3	3
6–10	6	9
11–15	12	21
16–20	9	30
21–25	2	32

The number of students who scored 1–5 is 3; the number who scored 1–10 is 9 (3 + 6); the number who scored 1–15 is 21 (9 + 12), and so on.

c) Draw the cumulative frequency curve on the axes below.   *(2 marks)*

Remember that you must plot the top value of the range in each class interval against the cumulative frequency. These are the coordinates (5, 3), (10, 9), (15, 21), (20, 30), (25, 32).

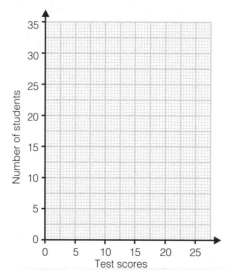
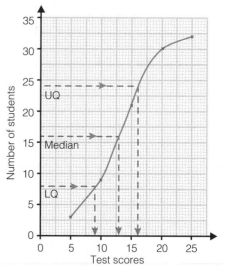

d) Using your cumulative frequency curve, find the median score and the interquartile range.   *(3 marks)*

The median score is 13.

The interquartile range is 16 − 9 = 7.

The top value in the cumulative frequency is 32. For the median, draw a horizontal line from the halfway point (16) to the curve, then vertically from this point to the *x*-axis. For the interquartile range, repeat this from the quarter (8) and three-quarter (24) points.

Practice questions

1. This dual bar chart shows the rainfall in millimetres in a particular area of Lancashire in two consecutive years.

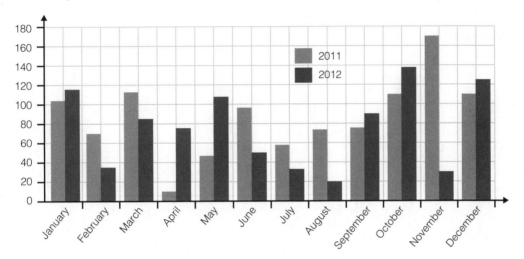

   a) In which month and year was the rainfall lowest? *(1 mark)*

   b) In which month and year was the rainfall highest? *(1 mark)*

   c) Which month shows the biggest difference from one year to the next? *(1 mark)*

   d) Which month shows the smallest difference from one year to the next? *(1 mark)*

2. Here is some information about the favourite subjects of students in Year 7.

Favourite subject	Year 7
Maths	6
English	8
Science	4
Geography	4
History	2
Total	**24**

   a) Draw a pie chart to represent this information. *(2 marks)*

   Compare your pie chart with this one for Year 8. There are 36 students in Year 8.

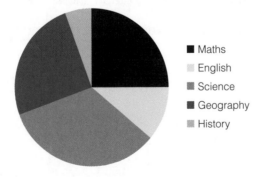

   ■ Maths
   ■ English
   ■ Science
   ■ Geography
   ■ History

   b) How many students prefer Science in Year 8? *(2 marks)*

c) Mary says that the same number of students prefer Maths in both groups?
Is she right? Give a reason for your answer.

(2 marks)

3. These scatter graphs show the relationships between SATs scores (*x*-axis) and three other variables (*y*-axis).

This graph shows the relationship between SATs scores and hours spent per week studying.

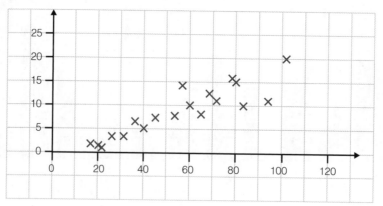

a) i) What type of correlation does this graph show? *(1 mark)*

ii) Describe the relationship between SATs scores and time spent studying. *(1 mark)*

iii) Draw a line of best fit on the graph above and estimate the SATs score for someone who spends $12\frac{1}{4}$ hours per week on study time. *(2 marks)*

This graph shows the relationship between SATs scores and hours spent per week playing on computer games.

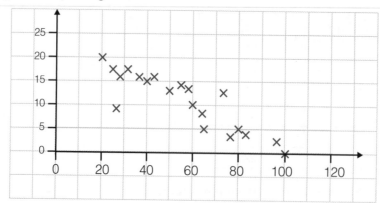

b) i) What type of correlation does this graph show? *(1 mark)*

ii) Describe the relationship between SATs scores and time spent on computer games. *(1 mark)*

c) This graph shows the relationship between SATs scores and shoe size.

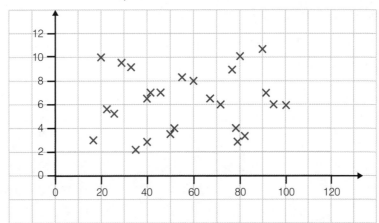

Describe the relationship between SATs scores and shoe size. *(1 mark)*

4. Here are the figures for the daily rainfall in mm in Padcombe Green over a particular fortnight earlier this year: 🔢

6   1   4   7   8   0   0   2   4   5   4   2   4   6

a) Find the mean. *(1 mark)*

b) Find the mode. *(1 mark)*

c) Find the median. *(1 mark)*

d) Find the range. *(1 mark)*

5. This table shows how many pets there are in each household in Troutcombe.

Number of pets	0	1	2	3	4	5	6
Frequency	9	13	8	4	2	1	0

a) Find the mean. *(1 mark)*

b) Find the mode. *(1 mark)*

c) Find the median. *(1 mark)*

d) Find the range. *(1 mark)*

6. Here is some numerical data:

31   23   17   2   21   6   31   36   19   37   7   46   27   31   11

a) Construct a stem-and-leaf diagram to represent this data. *(2 marks)*

b) Use your stem-and-leaf diagram to find each of the following:

i) The mean. *(1 mark)*

ii) The mode. *(1 mark)*

iii) The median. *(1 mark)*

iv) The range. *(1 mark)*

7. The following line graphs show the relationship between loss of uncultivated land and the population of lapwings.

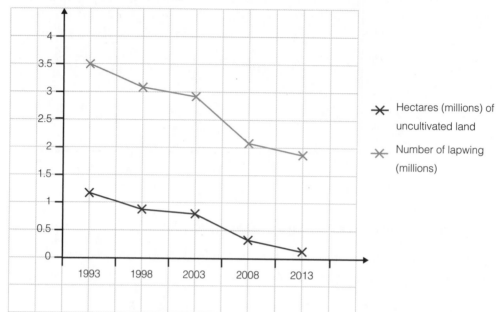

a) Estimate the amount of uncultivated land lost over the 20-year period. *(1 mark)*

b) How does the loss of uncultivated land relate to the lapwing population? *(1 mark)*

c) During which five-year period did the lapwing population fall the most? *(1 mark)*

d) Predict what might happen to the lapwing population in 10 years' time? *(1 mark)*

8. Find the midpoints of these groups of data.

a) *(1 mark)*

Scores	1–5	6–10	11–15	16–20	21–25	26–30
Midpoint						

b) *(1 mark)*

Height	$100 \leqslant h$ < 150	$150 \leqslant h$ < 200	$200 \leqslant h$ < 250	$250 \leqslant h$ < 300	$300 \leqslant h$ < 350	$350 \leqslant h$ < 400
Midpoint						

c) *(1 mark)*

Weight	$25 \leqslant w$ < 30	$30 \leqslant w$ < 35	$35 \leqslant w$ < 40	$40 \leqslant w$ < 45	$45 \leqslant w$ < 50	$50 \leqslant w$ < 55
Midpoint						

9. A vet has recorded the weights of dogs examined for annual health checks.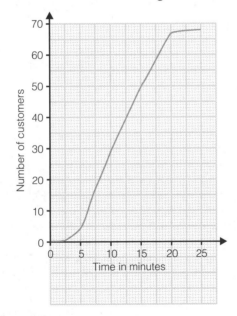

Weight (w kg)	Frequency (f)
$10 \leqslant w < 13$	8
$13 \leqslant w < 16$	10
$16 \leqslant w < 19$	16
$19 \leqslant w < 22$	8
$22 \leqslant w \leqslant 25$	9

a) Calculate the mean weight of the dogs. *(2 marks)*

b) Complete the cumulative frequency table. *(1 mark)*

Weight (w kg)	Frequency (f)	Cumulative frequency
$10 \leqslant w < 13$	8	
$13 \leqslant w < 16$	10	
$16 \leqslant w < 19$	16	
$19 \leqslant w < 22$	8	
$22 \leqslant w < 25$	9	

10. The cumulative frequency diagram shows the amount of time customers spent waiting for their meals in a restaurant after ordering.

a) What is the median waiting time? *(1 mark)*

b) What is the interquartile range? *(2 marks)*

c) Draw a box plot below the x-axis to show this data. *(2 marks)*

Chapter 1

1.1 Numbers, powers and roots

1. a) −18
    b) −6
    c) −1
    d) 10
2. a) True
    b) True
    c) False
    d) False
    e) True
    f) False
    g) True
3. a) ±5
    b) 4
    c) ±12
    d) −4
    e) 9
    f) 64
    g) 10 000
4. a) 4.2×10^7
    b) 6.32×10^5
    c) 3.21×10^{-2}
    d) 5.0047×10^4
    e) 6.4×10^{-5}
5. a) 6×10^{10}
    b) 6×10^{-6}
6. a) 1100
    b) 10 111
    c) 10 000

1.2 Fractions and decimals

1. $\dfrac{3}{28}, \dfrac{2}{7}, \dfrac{5}{14}, \dfrac{1}{2}$

2. a) $0.\dot{5}$
    b) 0.8
    c) $0.\dot{4}61\,53\dot{8}$
3. b
4. 8 students
5. 0.032, 0.046, 0.4694, 0.4702, 0.471
6. $\dfrac{7}{9}$
7.
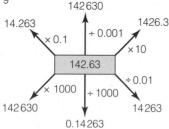

1.3 Percentages

1. a) $\dfrac{20}{100} = \dfrac{1}{5} = 0.2$

    b) $\dfrac{32}{100} = \dfrac{8}{25} = 0.32$

    c) $\dfrac{85}{100} = \dfrac{17}{20} = 0.85$

    d) $\dfrac{210}{100} = 2\dfrac{1}{10} = 2.1$

2. £100.80
3. $\dfrac{62}{80} \times 100\% = 77.5\%$

4. $\dfrac{25\,000}{165\,000} \times 100\% = 15.2\%$

5. $\dfrac{15\,000}{1.2} = 12\,500$

1.4 Ratio and proportion

1. £1.60
2. £30
3. 500 m
4.
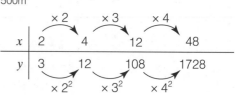

Chapter 2

2.1 Written and calculator methods

1. a) 876.8
    b) 1254
    c) 46
    d) 171 534
    e) 296.5
2. a) 14.45 (2 d.p.)
    b) 769.6 (1 d.p.)
    c) 7.052 (3 d.p.)
    d) $8\dfrac{1}{3}$ or $8.\dot{3}$

3. a) B since you do the multiplication part first.
    b) B since you do the multiplication part first.
    c) B since you work out $(9 + 1)^2$ first.
4. a) 0.057 067 271 = 0.057 07 (5 d.p.)
    b) She did not use brackets when putting the second line in her calculator.

2.2 Rounding and Estimating

1. a) 270
    b) 1270
    c) 42 960
    d) 2390
2. a) 47.37
    b) 21.43
    c) 15.37
3. a) 1200
    b) 0.0037

4. Since $8 \times 3 = 24$, which is not enough. Sandra will need 4 tins of paint.

5. **a)**
$$\approx \frac{30^2 + 100}{2 \times 5}$$
$$= \frac{900 + 100}{10}$$
$$= \frac{1000}{10}$$
$$= 100$$

b)
$$\approx \frac{500 + 200}{0.4}$$
$$= \frac{700}{0.4}$$
$$= \frac{7000}{4}$$
$$= 1750$$

6. **a)** $300 \times 40 = 12\,000$

b) Yes, Mr Johnson did make a mistake since his answer of 1226.4 is of the wrong magnitude.

Chapter 3
3.1 Symbols and Formulae

1. **a)** $4x - 8 + 3x - 3$
$$= 7x - 11$$

b) $n^2 + 2n + 1 - 2n - 4$
$$= n^2 - 3$$

2. **a)** $a^2 - 2ab + b^2$

b) $x^2 - x - 12$

c) $4a^2 + 4a - 3$

3. $60 = 2(20 + w)$
$$30 = 20 + w$$
$$w = 10$$

4. $F - 32 = \dfrac{9C}{5}$

$$C = \frac{5}{9}(F - 32)$$

5. $\dfrac{4a + 3b}{12}$

6. **a)** $2a^{10}$

b) $6a^4$

c) $4a^3 b$

d) $64a^4$

e) $\dfrac{16}{a^4}$ or $16a^{-4}$

7. **a)** $(x + 3)(x + 5)$

b) $(x + 5)(x - 2)$

c) $(x + 5)(x - 5)$

3.2 Equations and inequalities

1. **a)** $5x = 12 + 2$
$$x = 2.8$$

b) $4x = 18 - 2$
$$x = 4$$

c) $5x - 2x = 9 - 3$
$$3x = 6$$
$$x = 2$$

d) $6x - 2x = 15 + 1$
$$4x = 16$$
$$x = 4$$

e) $2x - 2 = 12x + 12$
$$-2 - 12 = 12x - 2x$$
$$-14 = 10x$$
$$x = -1.4$$

f) $7n + 11 = 39$
$$7n = 39 - 11$$
$$n = 4$$

2. **a)** $12x + 2 = 74$

b) $12x + 2 = 74$
$$12x = 72$$
$$x = \frac{72}{12}$$
$$x = 6, l = 34, w = 3$$

3. 3.3

4. $a = 2, b = 1$

5. **a)** $2x < 12$
$$x < 6$$

b) $5x \geqslant 21 - 1$
$$x \geqslant 4$$

c) $3 \leqslant 3x \leqslant 9$
$$1 \leqslant x \leqslant 3$$

Chapter 4
4.1 Sequences and functions

1. **a)**

Pattern number (n)	1	2	3	4	5	6
Perimeter (cm)	6	10	14	18	22	26

b) $4n + 2$

c) $4 \times 50 + 2 = 202$cm

2. **a)** 22, 25 (add 3 each time)

b) $\dfrac{1}{16}, \dfrac{1}{32}$ (multiply the denominator of the previous term by 2 each time)

c) 81, −243 (multiply the previous term by −3)

3. **a)** $T(n) = 2n + 3$

b) $T(n) = 3n + 4$

c) $T(n) = 2n^2$

4. **a)** 2, 5, 10, 17

b) 8, 6, 4, 2

c) −1, 6, 15, 26

4.2 Graphs of functions

1. **a)–b)**

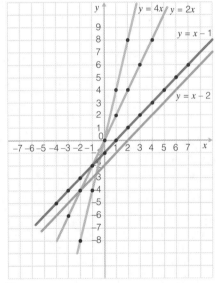

c) $y = 4x$ is steeper than $y = 2x$. They both pass through the origin.

d) See grid above

2. **a)** Gradient 4, intercept $(0, -1)$

b) Gradient −2, Intercept $(0, 3)$

c) Gradient 2, intercept $(0, 4)$

3. Graph A: $y = 3 - x^2$

Graph B: $y = 5 - x$

Graph C: $y = x^3$

4.3 Interpreting graphical information

1. **a)** 20mph

b) Car is stationary

c) 40mph

Answers to progress check questions

2. a) Point A represents the initial £50 charge for hiring the van.

  b) Gradient = 20. Hence £20 was charged per day.

  c) $C = 20d + 50$

3. A – Graph 2

  B – Graph 3

  C – Graph 1

  D – Graph 4

Chapter 5

5.1 Two and three-dimensional shapes

1.

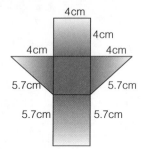

2.

 or

3. Yes congruent because SAS, i.e. two sides and the included angle are equal.

4. a)–b)

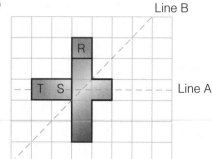

5. a)  **b)** **c)**

5.2 Angles, bearings and scale drawings

1. True

2. a) $a = 180° - 70° - 40°$

  $a = 70°$

  b) $x = 40°$

  $y = 140°$

  $z = 140°$

  c) $a = 55°$

  $b = 180° - 55° - 55°$

  $b = 70°$

  d) $x = 360° - 120° - 90° - 80°$

  $x = 70°$

3. $50\,000 \times 14 = 700\,000$ cm

  $= \dfrac{700\,000}{100\,000}$

  $= 7$ km

4. a) R from T = 070°

  b) R from P = 115°

5. Exterior angle $= \dfrac{360°}{5}$

  $= 72°$

  Interior angle $= 180° - 72°$

  $= 108°$

5.3 Pythagoras' theorem

1. a) $x = \sqrt{13.8^2 + 10.2^2}$

  $x = 17.2$ cm (3 s.f.)

  b) $x = \sqrt{25^2 - 15^2}$

  $x = 20.0$ cm (3 s.f.)

2. $h = \sqrt{5^2 - 1.75^2}$

  $h = 4.7$ cm

3. a) 10

  b) (4, 6)

5.4 Trigonometry in right-angled triangles

1. a) $x = 12 \times \tan 40°$

  $x = 10.1$ cm

  b) $x = \dfrac{9}{\cos 60°}$

  $x = 18$ cm

2. a) $\theta = \tan^{-1}\left(\dfrac{12}{25}\right)$

  $x = 26°$

  b) $\theta = \cos^{-1}\left(\dfrac{13}{28}\right)$

  $x = 62°$

3. 17.3 cm

4. a) 10.0 km (1 d.p.)

  b) 059° (nearest degree)

Chapter 6

6.1 Transformations and similarity

1. a)–d)

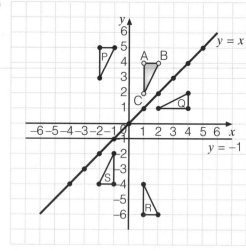

2. Each length should be three times the size of the original.

3. $\dfrac{x}{5} = \dfrac{14}{9}$

  $x = \dfrac{14}{9} \times 5$

  $x = 7.8$ cm (2 s.f.)

6.2 Constructions and loci

1.

In a test you would be expected to draw a scale diagram.

2. The angle must be drawn accurately to 40° ($\pm 1°$) and construction lines shown for the bisector.

3.

4.

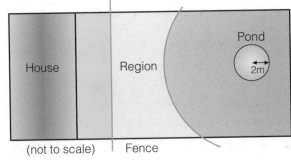

(not to scale)   Fence

Chapter 7

7.1 Units of measurement
1. $2 \times 2.2 = 4.4$lb
2. **a)** 23 hours and 25 minutes
   b) 0315
3. $5 \times 1000 = 5000$ml
4. $5 \times 2.5 = 12.5$cm
5. $V = 80.65$cm^3 (2 d.p.)

7.2 Area and perimeter of 2-D shapes
1. **a)** $A = \frac{1}{2} \times (a + b) \times h$

   $A = \frac{1}{2} \times (4.2 + 12.6) \times 8.1$

   $A = 68.0$cm^2 (3 s.f.)

   b) $A = b \times h$
   $A = 12 \times 5.3$
   $A = 63.6$cm^2

   c) $A = \pi \times r^2$
   $A = \pi \times 4.5^2$
   $A = 63.6$cm^2

   d) $A = b \times h$
   $A = 15 \times 8$
   $A = 120$ cm^2
   $A = \pi \times r^2$

   $A = \pi \times \dfrac{7.5^2}{2}$

   $A = 88.357$cm^2
   Total = 208cm^2 (3 s.f.)

2. Area of square = 10×10
   $= 100$cm^2
   Area of circle = $\pi \times r^2$
   $= \pi \times 5^2$
   $= 78.54$cm^2
   Shaded region = $100 - 78.54$
   $= 21.5$cm^2 (3 s.f.)

3. C
4. $A = \frac{1}{2} \times (a + b) \times h$

   $A = \frac{1}{2} \times (4.5 + 6) \times 4$

   $A = 21$m^2
   Cost = £35.99 × 21
   = £755.79

7.3 Volume of 3-D solids
1. **a)** $V = \frac{1}{2} \times b \times h \times l$

   $V = \frac{1}{2} \times 27.2 \times 6.5 \times 19.8$

   $V = 1750$cm^3 (3 s.f.)

b) $V = \pi \times r^2 \times h$
   $V = \pi \times 42.5^2 \times 10.6$
   $V = 60\,100$ cm^3 (3 s.f)

2. SA $= 2 \times (l \times h) + 2 \times (w \times h) + 2 \times (w \times l)$
   $= 2 \times (10 \times 6) + 2 \times (4 \times 6) + 2 \times (4 \times 10)$
   SA = 248cm^2

3. $h = \dfrac{v}{\pi r^2}$

   $h = \dfrac{2000}{\pi \times 5.6^2}$

   $h = 20.3$cm (3 s.f)

4. Volume scale factor = k^3
   $10 \times 3^3 = 270$cm^3

5. Area of cross-section = $4x^2$
   Volume $= 4x^2 \times 5x = 20x^3$

Chapter 8

8.1 The probability scale
1. **a)** very unlikely
   b) impossible
   c) likely

2. **a)** $\dfrac{6}{11}$

   b) $\dfrac{5}{11}$

   c) $\dfrac{11}{11} = 1$

   d) 0

3. **a)** $\dfrac{7}{11}$

   b) 0

   c) $\dfrac{4}{11}$

4. **a)** $\dfrac{15}{28}$

   b) $\dfrac{13}{28}$

   c) 0

5. 0.37

6. **a)** $\dfrac{3}{1000} = 0.003$

   b) $\dfrac{20}{1000} = 0.02$

   c) $\dfrac{21}{1000} = 0.021$

   d) $\dfrac{988}{1000} = 0.988$

7. 0.36

8. $1 - \dfrac{7}{9} = \dfrac{2}{9}$

8.2 Possible outcomes for two successive events
1. Mushroom and pineapple, pineapple and ham, mushroom and ham.

2.

		Dice 1				
	1	2	3	4	5	6
1	1	2	3	4	5	6
2	2	4	6	8	10	12
3	3	6	9	12	15	18
4	4	8	12	16	20	24
5	5	10	15	20	25	30
6	6	12	18	24	30	36

(Dice 2 labels the rows 1–6)

a) $\dfrac{2}{36} = \dfrac{1}{18}$

b) $\frac{11}{36}$

3. $0.8 \times 0.45 = 0.36$

4. a) $\frac{4}{87}$

b) $\frac{87}{112}$

5. a)

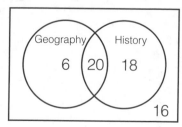

b) $x = 60 - 6 - 20 - 16$
$x = 18$

c) $20 + 6 = 26$

d) $\frac{16}{60} = \frac{4}{15}$

8.3 Estimating probability

1. $0.65 \times 200 = 130$

Therefore approximately 130 people will pass the test first time.

2. $\frac{58}{320} = \frac{29}{160}$

3. $\frac{4}{9} \times 1800 = 800$

Therefore approximately 800 students will get flu this winter.

4. a) The number 4 has come up far more times than would be expected, so it seems that the dice is more biased towards the 4.

b) Samuel could throw the dice more times, since the greater the number of times it is thrown the nearer to the theoretical probability.

Chapter 9

9.1 Identifying and selecting data

1. a) Secondary
b) Primary
c) Primary
d) Secondary
e) Secondary

2. This would be biased because students are more likely to be ill in the winter months than the summer months.

9.2 Collecting data

1. a) The groups overlap, for example which box would a 30-year-old person tick?

b) 'Sometimes', 'Occasionally' and 'Often' can mean different things to different people.

9.3 Organising data

1. a)

Score	Tally	Frequency
1–10	|||	3
11–20	||||| ||	7
21–30	|||||	5
31–40	|||||		6
41–50	||||| ||||	9

b) The 41–50 class interval.

2.

	Up to 20 miles	Over 20 miles	Total
Car	45	55	100
Train	120	30	150
Total	165	85	250

30 people had travelled more than 20 miles by train.

Chapter 10

10.1 Statistical diagrams

1. a) $\frac{140°}{360°} \times 720 = 280$ girls.

b) $\frac{70°}{360°} \times 720 = 140$ girls. Or since $140° = 280$ girls then $70° = 140$ girls.

2. a) Positive correlation – the better you did in Test 1, the better you did in Test 2.

b) The line of best fit should follow the direction of the data and have roughly the same number of points above it as below it.

3. There are no scales and Bar B is wider than Bar A.

10.2 Averages and range

1. a)

14	0 2
15	3 3 4 4
16	0 2 4 8
17	2 7
18	1

Key: 14 | 0 means 140cm

b) 41cm

c) 160cm

2. a) $\frac{\Sigma fx}{f} = \frac{617.5}{47}$
$= 13.1$cm (3 s.f.)

b) $10 \leqslant l < 15$

10.3 Cumulative frequency graphs

1. A cumulative frequency graph should be plotted with the points (0, 0), (5, 15), (10, 75), (15, 142), (20, 172), (25, 194), (30, 200).

2. a) 12 miles (approx.)

b) 9 miles (approx.)

Answers to practice questions

Chapter 1

1. **a)** One thousand; 1000   ↰ 4

   b) Three hundredths; 0.03 or $\frac{3}{100}$ ↰ 4

   c) 1×10^3 **(1 mark)** 3×10^{-2} **(1 mark)**   ↰ 12

   d) 742.7, 742.04, 741, 728.72, 724.8, 724.42   ↰ 18
   (2 marks if fully correct; 1 mark if two numbers are not in the correct order)

2. **a)** ≠   ↰ 19
   b) =   ↰ 19
   c) ≠   ↰ 19

3. **a)** >   ↰ 6
   b) >   ↰ 6
   c) >   ↰ 6

4. **a)** 31, 41, 61, 71, 101   ↰ 7

   b) $\frac{1}{13} = \frac{7}{91}$ **(2 marks)**   ↰ 7

   c) 231 is not a prime number. 2 + 3 + 1 = 6.
   Therefore it will divide by 3. It is also divisible by 7.
   Therefore it cannot be a prime number.   ↰ 11

5. **a)** 15m   ↰ 8
   b) 10m   ↰ 14
   c) The fishing net and the deck of the boat   ↰ 8/9
   d) 34m   ↰ 7

6. **a)**

1	24
2	12
3	8
4	6
5	■

   (1 mark for 12, 8 and 6 correctly placed in the table)

1	32
2	16
3	■
4	8

   (1 mark for 2, 4, 8 and 16 correctly placed in the table)

1	42
2	21
3	14
4	■
5	■
6	7

   (1 mark for 2, 3, 6, 7, 14 and 21 correctly placed in the table)   ↰ 7

   b) 2   ↰ 8

   c) i)

   24

   2 12

   2 6

   2 3

   $24 = 2^3 \times 3$ **(1 mark)**   ↰ 8

   42

   2 21

   3 7

   $42 = 2 \times 3 \times 7$ **(1 mark)**

   ii) $2^3 \times 3 \times 7 = 168$ **(1 mark)**   ↰ 9

7. **a)** 0.8   ↰ 17

   b) $0.2\dot{6}$ ↰ 17

   c) $\frac{76}{99}$ ↰ 18

8. **a)** $\frac{1}{6}$ ↰ 9

   b) $1\frac{1}{4}$ or $\frac{5}{4}$ ↰ 9

9. **a)** 0.125   ↰ 9

   b) $0.1\dot{3}$ ↰ 9

   c) $0.0\dot{9}$ ↰ 11

10. **a)** $\frac{11}{60}$ is beech. **(2 marks)**   ↰ 15

    b) $0.18\dot{3}$ ↰ 17

11. **a) i)** Both mums ate the same amount.   ↰ 16

    ii) Jane's mum ate $\frac{2}{5}$ of $\frac{5}{6}$ **(1 mark)**

    Imran's mum ate $\frac{5}{6}$ of $\frac{2}{5}$ **(1 mark)**

    The answer is the same. **(1 mark)**   ↰ 16

    b) $\frac{5}{6}$ of 30 = 25

    $\frac{2}{5}$ of 25 = 10

    $\frac{2}{5}$ of 30 = 12

    $\frac{5}{6}$ of 12 = 10

    or $\frac{2}{5} \times \frac{5}{6} = \frac{1}{3}$ (after cancelling)

    $\frac{1}{3}$ of 30 = 10

    (3 marks for either method)   ↰ 16

12. **a)** 6 hours   ↰ 30
    b) 8 carrots **(1 mark)** and 28 carrots **(1 mark)**   ↰ 28

13. Denmark is 750.25 ÷ 8.62 = £87.04
    Canada is 150.33 ÷ 1.72 = £87.50
    Spain is 105.5 ÷ 1.16 = £90.95
    Denmark would be cheapest.
    (1 mark for the correct individual totals and 1 mark for stating Denmark as the cheapest)   ↰ 29

14. If Mrs Shoesmith has paid 30%, she has 70% still to pay. To find 70% first find 10%, then multiply the answer by 7.
    10% of £800 = £80
    70% = £80 × 7 = £560   ↰ 23

15. a) 10% of £1800 = £180

5% of £1800 = £90

$2\frac{1}{2}$% of £1800 = £45

Mr Brown saves £45 each month ⊃ 23

b) $\dfrac{£280}{£1600}$

$= \dfrac{7}{40} \times 100$

= (after cancelling) $\dfrac{7}{2} \times 5 = 17\frac{1}{2}$

Mrs Brown saves $17\frac{1}{2}$% of her salary. ⊃ 23

16. a) i) $19\,600 \times 1.03 \times 1.03 \times 1.03$ ⊃ 26

ii) Each year Joan's new salary = £19 600 + 3%
of £19 600 = 1 + 0.03 = 1.03 **(1 mark)**
She receives this increase three times. **(1 mark)** ⊃ 26

b) i) $15\,000 \times 0.98 \times 0.98 \times 0.98$ ⊃ 26

ii) Joan's investment decreased by 2%. So she only
received 98% of her investment
= £15 000 – 2% of £15 000
= 1 – 0.02
= 0.98 **(1 mark)**
This happened in three consecutive
years. **(1 mark)** ⊃ 26

c) This tells us that Charlie's car depreciated in value by
14% in the first year (1 – 0.86 = 0.14 = 14%) **(1 mark)**
and by 30% in the second year (1 – 0.7 = 0.3 = 30%)
(1 mark) ⊃ 25/26

d) £146 000 ÷ 1.13 **(1 mark)**
or (£146 000 ÷ 113) × 100 **(1 mark)** ⊃ 27

17. a) 480 ÷ 20 = 24

24 × 1.5 **(1 mark)**

= 36ml **(1 mark)** ⊃ 29

b) 48.75 ÷ 1.5 = 32.5

32.5 × 20 **(1 mark)**

= 650kg **(1 mark)** ⊃ 29

18. a) 3 : 2 : 1 ⊃ 28

b) 3 + 2 + 1 = 6 parts
1 part = 7.2kg ÷ 6 = 1.2kg **(1 mark)**
Barley = 2 parts = 1.2 × 2 = 2.4kg
= 2400g **(1 mark)** ⊃ 28

19. $3\frac{3}{7} \times \frac{1}{2}$ or $3\frac{3}{7} \div 2 = 1\frac{5}{7}$ **(1 mark)**

$1\frac{1}{5} \div 1\frac{5}{7} = \frac{7}{10}$ cm **(1 mark)**

Or $1\frac{1}{5} \div 3\frac{3}{7} \times \frac{1}{2} = \frac{7}{10}$ cm **(2 marks)** ⊃ 16

Chapter 2

1. a) 75 999 + 15 950 = £91 949 ⊃ 40

b) Joe Tap = 42 × 12.50 = £525
Jill Washer = 96 × 5.25 = £504
Jason Watt = 38 × 13.75 = £522.50
Jamilla Spark = 112 × 4.75 = £532
**(1 mark for correctly calculating each pair of
quotations)**
The builder should choose Jill Washer and Jason Watt.
(1 mark) ⊃ 42

c) 504 + 522.50 = £1026.50 **(1 mark)**
15 950 – 1026.50 = £14 923.50 **(1 mark)** ⊃ 41

2. a) 45 × 38 = 1710 magazines ⊃ 42

b) The number of bundles which can be made from
2058 magazines ⊃ 42

c) 2058 ÷ 38 = 54.16 **(1 mark)**
32 more magazines are needed **(1 mark)** ⊃ 42/49

3. a) i) 60 000 ⊃ 47

ii) 0.079 ⊃ 47

b) $\dfrac{240}{0.8}$ **(1 mark)**

= 300 **(1 mark)** ⊃ 47

4. a) 4 ⊃ 43

b) 16 ⊃ 43

c) 88 ⊃ 43

d) 64 ⊃ 43

5. a) 234 ⊃ 45

b) 505 km per hour ⊃ 48

6. a) The largest area of the patio ⊃ 48

b) The smallest perimeter of the patio ⊃ 48

7. Lower limit: 100m **(1 mark)**
Upper limit: 104m **(1 mark)** ⊃ 48

8. Largest = 154 **(1 mark)**
Smallest = 145 **(1 mark)** ⊃ 45

9. $24.5 \leqslant x < 25.5$
$18.5 \leqslant y < 19.5$

a) Greatest value for $x - y = 25.5 - 18.5$ **(1 mark)**
= 7 **(1 mark)** ⊃ 48

b) Least value for $x \div y = 24.5 \div 19.5 = 1.2564102$ **(1 mark)**
= 1.26 **(1 mark)** ⊃ 48

10. a) Maria did not enter the brackets ⊃ 44

b) A display of 42.5 when calculating money means
£42.50 ⊃ 44

11. a) 240 ÷ 0.4
= 600 ⊃ 46

b) 246.12 **(1 mark)** ÷ 0.36
= 683.67 **(1 mark)** ⊃ 40/41/42

Chapter 3

1. a) $(3 \times 3^2) - 16 = 11$ ⊃ 63

b) $(3 \times 4 \times 5) + \dfrac{12}{3 \times 4} = 61$ ⊃ 63

2. a) 6(3x + 1) cm **(1 mark)**
18x + 6 **(1 mark)** ⊃ 59

b) $y + 2 + 2y + 4 + 4y - 3$ **(1 mark)**
7y + 3 **(1 mark)** ⊃ 58

c) £15s + £20c + £10b ⊃ 57

3. a) $2ab^2 + a^2b$ ⊃ 58

b) 22t + 3s ⊃ 59

c) 6y – 8sy + 8s ⊃ 60

d) 11q + 3p + 6pq + 3 ⊃ 59/60

4. a) $p = \dfrac{b}{2}$ ⊃ 64

b) $p = \dfrac{z - 4}{3}$ ⊃ 64

c) $p = \dfrac{2s}{q}$ ⊃ 64

d) $p = \sqrt{\dfrac{r}{3q}}$ ⊃ 64

5. a) $6x^2y^3$ ⊃ 62

b) $8x^7$ ⊃ 62

c) $6y^4$ ⊃ 62

6. a) $x^2 + 5x + 4$ ⊃ 60

b) $x^2 - 3x - 10$ ⊃ 60

c) $x^2 - 4x + 3$ ⊃ 60

d) $x^2 + 6x + 9$ ⊃ 60

e) $x^2 - 8x + 16$ ⊃ 60

7. a) 4x(2x + 3) ⊃ 60

b) 3ab(2 + 3b – a) ⊃ 60

c) (r – 8) (r + 8) ⊃ 61

d) (y + 6) (y + 1) ⊃ 61

(1 mark for each correct set of brackets)

8. a) $3p = -3$
   $p = -1$ ⟲65
   b) $q = 8 \times 4$
   $q = 32$ ⟲65
   c) $4x = x + 15$
   $3x = 15$
   $x = 5$ ⟲66
   d) $4x + 8 = 18x - 6$
   $4x + 14 = 18x$
   $14 = 14x$
   $x = 1$ ⟲66

9. a) $\dfrac{3x}{7} + 5 = x - 3$ ⟲67
   b) 14 ⟲67

10. a) $3x - 6 = 30$ ⟲67
   b) Mary – 12 **(1 mark)**
   Alex – 11
   James – 7 **(1 mark)** ⟲67

11. a) i) $4(2x) + 15 = 71$
   ii) 7km
   b) i) $2p + 35 = 67$
   ii) 16km ⟲67

12. a) $\dfrac{2x + 1}{3}$ ⟲61
   b) $\dfrac{x + 7}{21}$ ⟲61
   c) $\dfrac{3x}{4}$ **(1 mark)**
   $\dfrac{3xy}{4y}$ **(1 mark)** ⟲61
   d) $\dfrac{2xy}{3}$ or $\dfrac{2y}{3}$ **(1 mark)**
   $\dfrac{2y}{3}$ **(1 mark)** ⟲61

13. $4x + 2y = 46$ – Equation 1
   $3x - 2y = 10$ – Equation 2
   $4x + 3x + 2y + -2y = 56$
   (1 mark for adding the two equations)
   $7x = 56$
   $x = 8$ **(1 mark for correct value for x)**
   Substitute 8 for x in equation 1
   $32 + 2y = 46$
   $2y = 14$
   $y = 7$ **(1 mark for correct value for y)** ⟲68

14. $y = 6 - 4x$ ⟲64

15. $y + 4x = 6$ – Equation 1
   $3y - 2x = 4$ – Equation 2
   Make y the subject of equation 1:
   $y = 6 - 4x$
   Substitute $6 - 4x$ for y in equation 2:
   $3(6 - 4x) - 2x = 4$
   $18 - 12x - 2x = 4$
   $18 - 14x = 4$
   $-14x = -14$
   $x = 1$ **(1 mark for correct value for x)**
   So $y = 6 - 4$
   $y = 2$ **(1 mark for correct value for y)**
   (1 mark for showing the substitution clearly.) ⟲68

16. a) Joseph: $4a + 2c = £44 \times 3$
   Farrah: $3a + 3c = £41.25 \times 2$
   $12a + 6c = £132$ – Equation 1
   $6a + 6c = £82.50$ – Equation 2
   $6a = £49.50$
   $a = £8.25$
   Clear use of simultaneous equations **(1 mark)**
   £8.25 **(1 mark)** ⟲68

   b) Substitute £8.25 for a in equation 1.
   $£33 + 2c = £44$
   $2c = £11$
   $c = £5.50$
   £5.50 ⟲68
   c) £22 ⟲68

17. a) 3.5 **(1 mark for testing 3 and 4, 1 mark for testing 3.5, 1 mark for testing 3.45)** ⟲69
   b) 3.91 **(1 mark for testing 3 and 4, 1 mark for realising the number lies between 3.9 and 4.0, 1 mark for testing 3.95 and 1 mark for testing 3.915)** ⟲69

18. a) $-3, -2, -1, 0, 1, 2$ ⟲70
   b) ⟲70

   c)

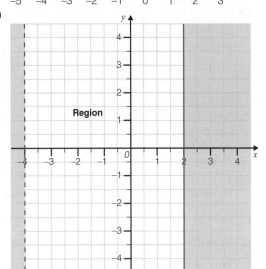

   (1 mark for dotted vertical line through −4; 1 mark for solid vertical line through +2; 1 mark for labelling the region) ⟲71

19. $x = -10$ **(2 marks)** ⟲66

Chapter 4

1. a) $(-2, 3)$ $(4, 3)$ ⟲82/84
   b) $(4, -5)$ $(4, 3)$ ⟲82/84
   c) $(-1, -1)$ $(4, 3)$ ⟲82/84
   d) $(4, -5)$ $(-2, 3)$ ⟲82/84

2. a) P and Q **(1 mark for each)** ⟲82
   b) R ⟲82
   c) S and T **(1 mark for each)** ⟲82
   d) P and Q **(1 mark for each)** ⟲82

3. a)

(1 mark for each correctly marked axis. 1 mark for a correctly plotted line) ⟲88
   b) See graph above. **(1 mark for each correct conversion)** ⟲88

4. a) B and F **(1 mark for each)**   ⤴ 89
    b) C   ⤴ 89
    c) E, 15 minutes.   ⤴ 89
    d) She returned home **(1 mark)**. She travelled by car or bus
        (1 mark).   ⤴ 89

5. a) £300   ⤴ 88
    b) £440   ⤴ 88
    c) $\dfrac{2}{3}$ ⤴ 88
    d) Gordon's wage = $\dfrac{2x}{3} + 300$ ⤴ 88
    e) £580   ⤴ 88
    f) 630 boxes   ⤴ 88

6.

Sequence					n^{th} term
6	11	16	21	26	$5n + 1$
8	17	26	35	44	$9n - 1$
5	8	11	14	17	$3n + 2$
2	8	18	32	50	$2n^2$
1	6	13	22	33	$n^2 + 2n - 2$

(1 mark for each correct answer)   ⤴ 79

7. a) $3n + 1$ ⤴ 80
    b) 121 rings   ⤴ 80

8. a)

$x =$	−5	−4	−3	−2	−1	0	1	2	3
x^2	25	16	9	4	1	0	1	4	9
$+2x$	−10	−8	−6	−4	−2	0	2	4	6
	−2	−2	−2	−2	−2	−2	−2	−2	−2
$y =$	13	6	1	−2	−3	−2	1	6	13

**(4 marks for all y-values correctly calculated; 3 marks
if at least seven values correctly calculated; 2 marks
if at least five values correctly calculated; 1 mark if at
least three values correctly calculated)**   ⤴ 86

    b)
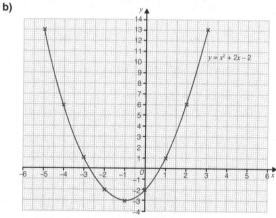

$y = x^2 + 2x - 2$

(2 marks for a fully correct graph)
    c) $x = 0.8$, $x = -2.7$ **(1 mark for each correct value)**   ⤴ 86

Chapter 5

1. First row completed as follows: Equilateral triangle
    Second row completed as follows: 4, 1, 1
    Third row completed as follows: Trapezium
    Fourth row completed as follows: 7, 7, 7, 7
    Fifth row completed as follows: 0
    **(4 marks for all correct; otherwise, 1 mark for every three
    correct answers)**   ⤴ 98/99

2. A – isosceles, right-angled
    B – equilateral
    C – isosceles
    D – scalene
    E – scalene, right-angled
    **(3 marks for all correct; otherwise, 1 mark for every three
    correct answers)**   ⤴ 98

3. A and B (SAS)   ⤴ 103

4. AO – radius
    DE – chord
    FG – perpendicular bisector
    HDI – tangent
    CA (curved) – arc
    BC – diameter
    **(3 marks for all correct; otherwise, 1 mark for each pair of
    correct answers)**   ⤴ 99

5. a) 72°   ⤴ 106
    b) 5   ⤴ 106
    c) pentagon   ⤴ 106

6. a) D is correct   ⤴ 102
    b) A is not a mirror image **(1 mark)**
        B is not level **(1 mark)**
        C is too far away from the central line **(1 mark)**   ⤴ 102

7. 3 – one
    8 – two
    609 – two
    909 – one
    69 – two
    **(3 marks for all correct; otherwise, 1 mark for each pair of
    correct answers)**   ⤴ 102

8.

Front Elevation   Side Elevation
(1 mark for each correct elevation)   ⤴ 101

9. a) i) 215°   ⤴ 105
        ii) Sum of x and y = 145° (angles of a triangle)
            (1 mark)
            $x + a + y + b = 360°$
            $360 - 145 = 215°$ **(1 mark)**   ⤴ 105
    b) $x = 65°$ **(1 mark)**
        $y = 57°$ **(1 mark)**   ⤴ 106

10. a) 39m   ⤴ 115/116
    b) ZB = 29m **(1 mark)**
        ZA = 31m **(1 mark)**
        Ramp = 41m **(1 mark)**   ⤴ 111
    c) 40°   ⤴ 115/116

11. Lawn: 8m × 4m
    Patio: 6m × 4m
    Flower bed: 14m × 2m
    (1 mark for each correct answer)   ⤴ 109

12. 260° (360 – 45 – 55)   ⤴ 108/109

Chapter 6

1. a) $x = -1$ ⤵ 124
   b) ⤵ 124

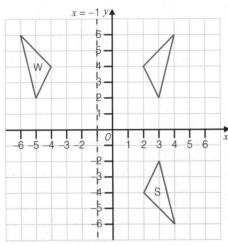

2. a) Translation **(1 mark)** by a vector of $\begin{pmatrix} 1 \\ -4 \end{pmatrix}$ **(1 mark)** ⤵ 125

   b) Triangle D should be plotted with the coordinates (6, 6), (6, 8) and (8, 6). ⤵ 125

   c) $\begin{pmatrix} 6 \\ -4 \end{pmatrix}$ ⤵ 125

3. a) 90° anti-clockwise **(1 mark)** about (0, 0) **(1 mark)** ⤵ 125
   b)

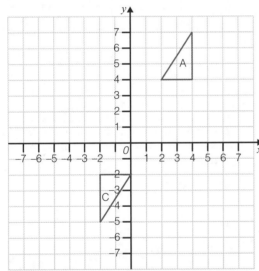

   (1 mark for correct orientation, 1 mark for correct position) ⤵ 125

4. a) $1\frac{1}{2}$ ⤵ 126
   b) 9cm ⤵ 126

5. a) A reflection **(1 mark)** in the line $x = 1$ **(1 mark)** ⤵ 124
   b) A rotation **(1 mark)** 180° clockwise **(1 mark)** around (−5, 1) **(1 mark)** ⤵ 125
   c) A translation **(1 mark)** by vector $\begin{pmatrix} -4 \\ 6 \end{pmatrix}$ **(1 mark)** ⤵ 125
   d) An enlargement **(1 mark)**, scale factor $\frac{1}{2}$ **(1 mark)** around (1, 3) **(1 mark)** ⤵ 126

6. 8.3cm ⤵ 127
7. $a = 2m$ **(1 mark)**, $b = 2.25m$ **(1 mark)** ⤵ 128

8.

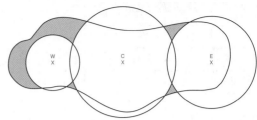

Circle around Western mast with a radius of 1.5cm **(1 mark)**
Circle around Eastern mast with a radius of 2.5cm **(1 mark)**
Circle around central mast with a radius of 3cm **(1 mark)** ⤵ 130

Chapter 7

1. a) 22 minutes ⤵ 141
   b) 13.23 ⤵ 141
   c) 42 minutes ⤵ 141
   d) 21 minutes ⤵ 141
   e) D = 20 × 0.75 **(1 mark)**
      = 15km **(1 mark)** ⤵ 142
   f) S = 15 ÷ 1.1 **(1 mark)**
      = 14 kilometres per hour **(1 mark for correct, rounded answer)** ⤵ 142

2. a) Flag = 10 000cm² in total.
      9000cm² will be green.
      $= \frac{9}{10}$ ⤵ 144
   b) Area of flag = 646cm²
      Area of circle = 63.6cm²
      646 − 63.6 = 582.4cm² is white ⤵ 146
   c) Area of red = 24cm²
      $24 = \frac{1}{2}(b \times h) = \frac{1}{2}(8 \times h)$
      $h = 6cm$
      Area of green = 18cm²
      $18 = \frac{1}{2}(b \times h)$
      $= \frac{1}{2}(b \times 6)$ (height is same as red triangle, 6cm)
      $b = 6cm$
      If base of green is 6cm then base of yellow = 8 − 6
      = 2cm ⤵ 144

3. a) Width of garage = 25 − 18 = 7m **(1 mark)**
      The area of the house can be split into two rectangles, one measuring 18m by 15m = 270m², the other 7m by unknown length.
      Total area of house = 312m²
      Therefore, the area of smaller rectangle = 312m² − 270m²
      = 42m².
      Unknown length = 42m² ÷ 7m = 6m
      Length of garage = 15m − 6m = 9m **(1 mark)**
      Dimensions of garage = 9m x 7m ⤵ 144/145
   b) Area of whole = area of lawn + area of quadrant
      Area of lawn = area of whole − area of quadrant
      Area of quadrant = $(\pi r^2) \div 4 = (3.142 \times 25) \div 4 =$ 19.6375 (19.6m² to 3 s.f.) **(1 mark)**
      Area of whole = 32m × 5m = 160m²
      Area of lawn = 160 − 19.6 = 140.4m² **(1 mark)** ⤵ 146
   c) $\frac{1}{2}(12 + 16) \times 10$ **(1 mark)**
      = 14 × 10 = 140m² **(1 mark)** ⤵ 144
   d) 140m² = 1 400 000cm²
      Number of flags = area of trapezium ÷ area of one flag
      = 1 400 000 ÷ 900
      = 1555.5555
      = 1556 flags ⤵ 144/147

4.

kg	g	kg	g	
7.04kg	7040g	7kg	40g	**(1 mark)**
7.004kg	7004g	7kg	4g	**(1 mark)**
7.4kg	7400g	7kg	400g	**(1 mark)**
0.07kg	70g	0kg	70g	**(1 mark)** ↺ 139

5. $2cm^2$ ——— $200mm^2$ **(1 mark)**
$2m^2$ ——— $20000cm^2$ **(1 mark)**
$2km^2$ ——— $2000000cm^2$ or $2m^2$ ——— $2000000mm^2$
(1 mark) ↺ 147

6. $14 \times 8 = 112cm^2$ ↺ 147

7. a) $18 = 2 \times 3 \times 3$
So depth = 3m ↺ 148

b) Volume of ice = surface area × thickness
$7500cm^3 = (300 \times 50) \times$ thickness
$7500 = 15000 \times$ thickness
Thickness of the ice = $7500 \div 15000$
= 0.5cm ↺ 148

c) $1m^3 = 1000$ litres
Tank holds 200 litres = $\frac{200}{1000} = \frac{1}{5}m^3$
$1m^3 = 100cm \times 100cm \times 100cm = 1000000cm^3$
200 litres = $\frac{1}{5}m^3 = 1000000 \div 5 = 200000cm^3$
Volume = length × width × depth
$200000 =$ length $\times 250 \times 10 =$ length $\times 2500$
Length = $200000 \div 2500$
= 80cm ↺ 148

8. a) $35 \div 7 = 5, 25 \div 5 = 5, 10 \div 2 = 5$ **(1 mark)**
$5 \times 5 \times 5$ **(1 mark)** = 125 boxes **(1 mark)** ↺ 148

b) Place 35cm edge along the width of the film **(1 mark)**
Minimum length of film = 70cm **(1 mark)** ↺ 148

9. a) $\frac{1}{2}(4 \times 4) = 8$
$8 \times 14 = 112cm^3$ ↺ 148

b) $D = 901.6 \div 112$ **(1 mark)**
= 8.05 grams per cm^3 **(1 mark)** ↺ 148

Chapter 8

1. a) The two events are mutually exclusive because they cannot happen at the same time. ↺ 159

b) Events are exhaustive if their probabilities add up to 1.
P(Spearmint) = $\frac{5}{15} = \frac{1}{3}$
P(LS) = $\frac{10}{15} = \frac{2}{3}$
$\frac{1}{3} + \frac{2}{3} = 1$ ↺ 158

2. a)

SU				
SO	UO			
SL	UL	OL		
ST	UT	OT	LT	
SM	UM	OM	LM	TM

(S – Superstars; U – United; O – Owls; L – Lightning; T – Thunder; M – Marvels)
(1 mark for an organised list; 1 mark for correct outcomes) ↺ 162

b)

	Probability of name being drawn	Description
Marvels	$\frac{1}{6}$	Possible
Owls	$\frac{1}{5}$	Possible
United	$\frac{1}{4}$	Possible
Lightning	$\frac{1}{3}$	Possible
Superstars	$\frac{1}{2}$	Even chance
Thunder	$\frac{1}{1}$	Certain

(1 mark for each correct column) ↺ 159/161

c) $\frac{1}{15}$

3. a) $0.18 + 0.02 = 0.2$ ↺ 160/163

b) $0.6 = 90$, so $0.1 = 15$
Total = 150 ↺ 160

c) $0.2 + 0.02 = 0.22$ ↺ 160

4. a)

	20p	21	22	25	30	40
	10p	11	12	15	20	30
	5p	6	7	10	15	25
Lily's choice	2p	3	4	7	12	22
	1p	2	3	6	11	21
		1p	2p	5p	10p	20p
				Ellie's choice		

(3 marks if fully correct; 2 marks for 20–24 of the total values correct; 1 mark for 15–19 of the total values correct) ↺ 162

b) $\frac{17}{25}$ ↺ 162

c) $\frac{1}{5}$ ↺ 162

d) P (odd total) = $\frac{12}{25}$
P (even total) = $\frac{13}{25}$
Theoretically, Lily is more likely to win overall. ↺ 162

5. a)

	4	4, 1	4, 2	4, 3	4, 4
	3	3, 1	3, 2	3, 3	3, 4
	2	2, 1	2, 2	2, 3	2, 4
Anil's choice	2	2, 1	2, 2	2, 3	2, 4
	1	1, 1	1, 2	1, 3	1, 4
	1	1, 1	1, 2	1, 3	1, 4
		1	2	3	4
			Josh's choice		

(3 marks if fully correct; 2 marks for 17–21 of the total values correct; 1 mark for 12–16 of the total values correct) ↺ 162

b) $\frac{1}{4}$ ↺ 162

c) Least likely = 4, 4
Most likely = 2, 1 (1, 2) ↺ 162

d) 2, 2, 2, 1, 3, 4 ↺ 159

6. a) 25 ↺ 159/160

b) $\frac{1}{3}$ ↺ 159/160

7. a) ↺ 164/165

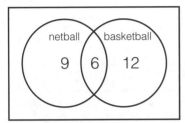

b) 3 ↺ 164/165

c) $\frac{3}{30} = \frac{1}{10}$ ↺ 164/165

8. Geography = 0.21, History = 0.13, Science = 0.21,
Art = 0.25, Sport = 0.2 **(2 marks if all correct; 1 mark for
at least two correct)** ↺ 159/160

9. a) 1st attempt = 0.05, 2nd attempt = 0.3,
3rd attempt = 0.27, 4th attempt = 0.19
**(2 marks if all correct; 1 mark for
at least two correct)** ↺ 169/170

b) The 4th attempt because this attempt had
most trials. ↺ 171

10. a) $\frac{1}{4}$ ↺ 159

b) 48 ↺ 28

c) yellow (stays at 12) ↺ 159

11. a) $\frac{1}{6} \times \frac{1}{6} = \frac{1}{36}$ ↺ 166

b) $\frac{5}{6} \times \frac{5}{6} = \frac{25}{36}$ ↺ 166

c) $\frac{1}{6} \times \frac{5}{6} = \frac{5}{36}$ ↺ 166

d) $\left(\frac{1}{6} \times \frac{5}{6}\right) + \left(\frac{5}{6} \times \frac{1}{6}\right) = \frac{10}{36} = \frac{5}{18}$

**(1 mark for $\frac{1}{6} \times \frac{5}{6}$ or $\frac{5}{6} \times \frac{1}{6}$ and 1 mark for correct
answer)** ↺ 166

Chapter 9

1. a) i) Secondary
ii) Primary
iii) Primary
iv) Secondary ↺ 178

b) ii) It measures time ↺ 178

2. a) Too wordy ↺ 181

b) Too sensitive (a person may not want to admit to having
done something wrong) ↺ 181

c) Too ambiguous (everyone's interpretation of frequently,
regularly and rarely is different) ↺ 181

d) Too biased (it leads person to agree)
**(2 marks for all four correct; 1 mark for at least two
correct)** ↺ 181

3. a) Too many variables to deal with **(1 mark)**
Limited time of survey will not give a true picture
(1 mark) ↺ 180/181

b) Outside a swimming pool will give a biased picture
(1 mark)
Observation sheet does not include men and women
(1 mark) ↺ 180/181

c) Horses kept in different environments will alter the
results. ↺ 180/181

4. a)

Time	Tally	Frequency
$24 \leqslant t < 27$	卌	5
$27 \leqslant t < 30$	卌 ||	7
$30 \leqslant t < 33$	卌 卌 |	11
$33 \leqslant t < 36$	卌 ||	7

**(2 marks for correct tally column; 1 mark for correct
frequency column)** ↺ 182

b) 5 ↺ 182

c) $\frac{1}{6}$ ↺ 182

5. a)

	RSPB member	Non-member	Total
First visit	45	**165**	210
Visited before	**306**	54	**360**
Total	**351**	**219**	570

**(3 marks for a fully correct two-way table; 2 marks for
finding at least three of the missing values; 1 mark
for finding at least two of the missing values)** ↺ 183

b) $\frac{306}{360} \times 100 = 85\%$ ↺ 183

c) 38% ↺ 183

Chapter 10

1. a) April 2011 ↺ 188

b) November 2011 ↺ 188

c) November ↺ 188

d) January ↺ 188

2. a) The pie chart should be constructed as follows:

Maths = $\frac{6}{24} = \frac{1}{4} = 90°$

English = $\frac{8}{24} = \frac{1}{3} = 120°$

Science = $\frac{4}{24} = \frac{1}{6} = 60°$

Geography = $\frac{4}{24} = \frac{1}{6} = 60°$

History = $\frac{2}{24} = \frac{1}{12} = 30°$

**(2 marks for a fully correct pie chart; 1 mark
if the angle for one subject has been wrongly
calculated)** ↺ 190

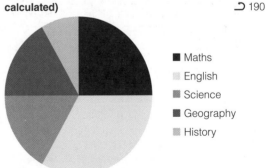

b) 12 **(2 marks) (1 mark for stating 120°)** ↺ 190

c) Mary is wrong **(1 mark)**
Both charts show 90° for Maths.
90° = one quarter of 360°
However, Year 7 = $\frac{1}{4}$ of 24 = 6 and Year 8 = $\frac{1}{4}$ of
36 = 9 **(1 mark)** ↩ 190

3. a) i) Positive
   ii) The longer the study time the higher the SATs score.
   iii) 70 **(1 mark)**

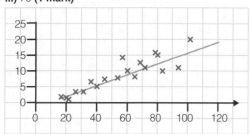

(1 mark) ↩ 193

b) i) Negative
   ii) The longer the time spent on computer games the
       lower the SATs score. ↩ 193
c) There is no correlation between someone's shoe size
   and their SATs score. ↩ 193

4. a) 53 ÷ 14 = 3.79 ↩ 196
   b) 4 ↩ 196
   c) 4 ↩ 196
   d) 8 − 0 = 8 ↩ 196

5. a) 1.5 ↩ 197
   b) 1 ↩ 197
   c) 1 ↩ 197
   d) 5 − 0 = 5 ↩ 197

6. a)

0	2	6	7		
1	1	7	9		
2	1	3	7		
3	1	1	1	6	7
4	6				

Key: 2 | 3 = 23
(1 mark for correct diagram and 1 mark for key) ↩ 198

b) i) 23
   ii) 31
   iii) 23
   iv) 46 − 2 = 44 ↩ 198

7. a) Approximately 1 million hectares ↩ 192
   b) As the amount of uncultivated land has decreased so
       has the lapwing population. ↩ 192
   c) 2003–2008 ↩ 192
   d) Any suitable answer, e.g.: If the land loss continues the
       lapwing population could be wiped out. ↩ 192

8. a) 3, 8, 13, 18, 23, 28 ↩ 198
   b) 125, 175, 225, 275, 325, 375 ↩ 198
   c) 27.5, 32.5, 37.5, 42.5, 47.5, 52.5 ↩ 198

9. a)

Weight (w kg)	Frequency (f)	Midpoint	Midpoint × frequency
$10 \leqslant w < 13$	8	11.5	92
$13 \leqslant w < 16$	10	14.5	145
$16 \leqslant w < 19$	16	17.5	280
$19 \leqslant w < 22$	8	20.5	164
$22 \leqslant w < 25$	9	23.5	211.5
	51		892.5

892.5 **(1 mark)** ÷ 51 = 17.5 **(1 mark)** ↩ 198
                                                    ↩ 200

b)

Weight (w kg)	Frequency (f)	Cumulative frequency
$10 \leqslant w < 13$	8	8
$13 \leqslant w < 16$	10	18
$16 \leqslant w < 19$	16	34
$19 \leqslant w < 22$	8	42
$22 \leqslant w < 25$	9	51

10. a) 11 minutes ↩ 201

b) $15\frac{1}{2} - 7\frac{1}{2} = 8$ minutes

**(2 marks if fully correct; 1 mark if $15\frac{1}{2}$ or
$7\frac{1}{2}$ seen)** ↩ 201

c)

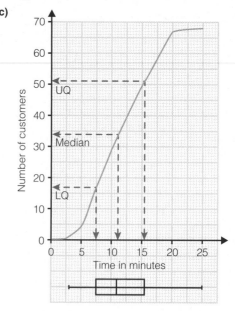

(2 marks if fully correct; 1 mark if one error) ↩ 202

Mixed test-style questions

These questions should be answered without the use of a calculator.

1. New Year's Day is celebrated at different times across the world. The following table shows the time differences between various places and the United Kingdom.

City	Country	Time difference (hours)
Christmas Island	Kiribati	+14
Fakaofo	Tokelau	+13
Funafuti	Tuvalu	+12
Brisbane	Australia	+10
Tokyo	Japan	+9
Athens	Greece	+2
London	United Kingdom	0
Lima	Peru	−5
San Francisco	USA	−8
Papeete	French Polynesia	−10
Alofi	Niue	−11
Baker Island	US territory	−12

a) Christmas Island is the first place in the world to celebrate the New Year.

When it is 00:01 on 1st January 2020 in Christmas Island, what will the time and date be in London? Give your answer in 24-hour clock.

(1 mark)

b) What is the time difference between Brisbane and San Francisco?

_____ hours *(1 mark)*

c) Give the names of two places with a time difference of one day.

(1 mark)

223

2. Look at these three spinners.

Spinner A   **Spinner B**   **Spinner C**

Give the probability of the spinner landing on a 1 on:

a) Spinner A

(1 mark)

b) Spinner B

(1 mark)

c) Spinner C

(1 mark)

3. Find the value of x in these equations:

a) $\dfrac{x}{9} = 3$

$x =$ _____ *(1 mark)*

b) $3(4x - 6) = 2(3x + 3)$

c) $3(4x - 6) = 2(3x + 3)$

$x =$ _____ *(1 mark)*

4. The following table gives information about three rectangles.

Complete the missing information.

Length (cm)	Width (cm)	Area (cm²)	Perimeter (m)
5	3		
	4	24	
		18	18

(3 marks)

5. Are these fractions equivalent?

$$\frac{8}{12} = \frac{14}{42}$$

Tick Yes or No.

☐ Yes   ☐ No

Explain your choice.

(1 mark)

6. **a)** Last year my apple trees produced 400kg of apples. This year they produced 12% more.

How many kilograms of apples did I get this year?

_____ kg *(1 mark)*

b) Last year my pear trees produced 300kg of fruit. This year they produced 12% less.

How much fruit did they produce this year?

_____ kg *(1 mark)*

c) After an increase of 22% on last year, my plum trees produced 366kg of fruit this year.

How much fruit did they produce last year?

_____ kg *(1 mark)*

7. The following graph shows the demise of UK grown food between 2000 and 2011.

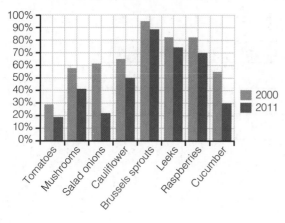

a) Which fruit/vegetable shows the greatest difference between 2000 and 2011?

(1 mark)

b) Which fruit/vegetable shows the least difference between 2000 and 2011?

(1 mark)

c) Complete this sentence:

_____ per cent of the cucumbers eaten in the UK in 2011 were
grown abroad.   *(1 mark)*

8. In a box of chocolate biscuits, 21 are covered with white chocolate, 28 with plain chocolate and 42 with milk chocolate.

a) Simplify these proportions to their lowest terms.

(1 mark)

b) How many plain biscuits would you find in a box of 156 biscuits?

(2 marks)

9. McGregory's drive-through takeaway is open from 7am until 11pm. Yesterday, the takeaway sold 564 meals to 257 customers.

a) $564 \div 257 = 2.195$

Complete this sentence:

 This calculation tells us that each customer bought about

_____ *(1 mark)*

b) $257 \div 16 = 18.357$

Complete this sentence:

 This calculation tells us that on average about _____

_____ *(1 mark)*

10.

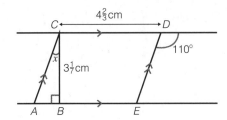

a) What is the value of angle x? Explain your answer.

$x =$ _____ *(2 marks)*

b) The area of the triangle ABC is 2.75cm².

What is the length of AB?

_____ cm *(2 marks)*

c) What is the area of the parallelogram $ACDE$?

_____ cm² *(1 mark)*

11. Complete the values of the missing terms in the table below.

Term 1	Term 2	Term 3	Term 4	Term n
7	9	11	13	$2n + 5$
13	11	9	7	
				$7n - 3$
				$3 - 7n$

(3 marks)

12. A linear graph $y = 2x - 4$ is drawn.

    a) Which two of the following points does it pass through?

    (2, 2)   (4, 4)   (0, 2)   (5, 6)   (2, −4)

    (1 mark)

    b) Which of the following lines would be parallel to $y = 2x - 4$?

    $2x + y = 4$ $y = 4 - 2x$ $2x - y = 8$ $2x = 4 - y$

    (1 mark)

13. The stem and leaf diagram shows the marks gained by pupils in a History test.

    ```
    1 | 2  5  6  9
    2 | 2  3  7  7  7  9
    3 | 4  4  6  7  9  9
    4 | 3  5  8  8
    ```
    Key: **1 | 2** = 12 marks

    a) What is the range? Show your working.

    Range = _____ *(1 mark)*

    b) What is the mean score for the class? Show your working.

    (2 marks)

A calculator can be used to answer these questions.

14. Find the answers to these.

 a) Add 8926 hundredths to 763 tenths.

(1 mark)

 b) Find the difference between 28 739 thousandths and 8400 hundredths.

(1 mark)

15. Jim and Polly have forgotten the four-digit lock code for the gate of their allotment.

 Jim can remember that the two outer numbers are square numbers and when put together as a 2-digit number make another square number.

 Polly can remember that the two inner numbers have a difference of 2 and make a prime number whichever way they are combined.

 What are the two combinations Jim and Polly tried? None of the numbers is a zero.

(2 marks)

16. Give the values of the following when $x = 5$ and $y = 7$.

 a) xy^2

(1 mark)

 b) $4(x + y) - 3(x - y)$

(1 mark)

229

17. A survey of favourite holiday destinations was carried out in Years 7, 8 and 9. The students could choose from Waltland (WL), Middle Parks (MP), Ludoland (LL) or Jungletrek (JT). The results were put into a two-way table.

	Year 7	Year 8	Year 9	Total
WL	9	6	9	
MP	10	7	7	
LL	9	12	11	
JT	2	3	5	
Total				

a) Complete the total boxes in the table. *(1 mark)*

b) Jane says that Middle Parks is equally popular in Year 8 and Year 9.

Is she correct? Tick Yes or No

 ☐ Yes ☐ No

Explain your answer.

(1 mark)

c) Waleed says that the probability of any random student choosing Waltland or Middle Parks is about 25%. Is he correct? Tick Yes or No

 ☐ Yes ☐ No

Explain your answer.

(1 mark)

18. Solve these simultaneous equations.

$4x - 4y = 8$

$x + 2y = 14$

$x =$ _____ $y =$ _____ *(2 marks)*

19. Anuja needs to reinforce the fencing around her horses' field with fencing tape. Fencing tape comes in rolls of 650cm.

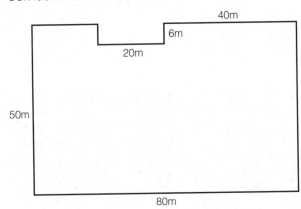

a) Anuja tells her dad that she will need 41 rolls. Her dad thinks she will need 42.

Who is correct? Explain your answer.

(1 mark)

b) Anuja remembers that she will not require tape to go across her gate, which is 420cm wide. She is sure that she will need 41 rolls but her dad thinks she will need 42.

Who is correct? Explain your decision.

(1 mark)

c) Fertiliser is spread at the rate of 200g per m². It is packed in bags containing 10kg.

How many bags of fertiliser should Anuja order?

_____ bags *(2 marks)*

20. Mr and Mrs Ahmed are planning a holiday with their two children to Ludoland. They need transport to Ludoland, two nights' accommodation and they will spend two days in Ludoland itself. They want to take the cheapest option.

From the information given below, calculate which will be the cheapest package. Show your working.

		ADULT	CHILD	EXTRA INFORMATION
TRAVEL	**FERRY**	£193.45	£98.65	Additional fuel costs £150
	AIR	£230.50	£75	Additional fuel costs £40 Parking costs £14
ACCOMMODATION	**FAMILY ROOM**	£150 per night		Breakfast included
	HOTEL ROOMS	£90 per night for 2 adults	£90 per night for 2 children	Second night half price. Breakfast £7 per adult per day. Children free.
ENTRY TO LUDOLAND		Family pass £86 per day		25% reduction on second day
		£27 per adult per day	£16 per child per day	

Tick your choices.

TRANSPORT: Ferry ☐ Air ☐

ACCOMMODATION: Family room ☐ Hotel rooms ☐

ENTRY COSTS: Family pass ☐ Daily tickets ☐ (3 marks)

21. Tom services fire extinguishers for a living. He charges £5 for every appliance serviced. The graph illustrates his charges.

a) What does Tom charge before he services any appliances?

£ _____ *(1 mark)*

b) What is the gradient of the line?

(1 mark)

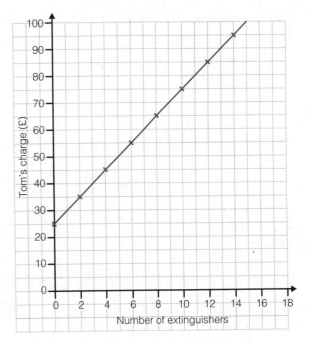

c) Write down the equation for Tom's charge in terms of the number of fire extinguishers he services.

(1 mark)

d) Use the equation to calculate what Tom would charge to service 95 fire extinguishers.

£ _____ *(1 mark)*

22. Look at this diagram.

a) Describe the transformation which takes square B to C.

(1 mark)

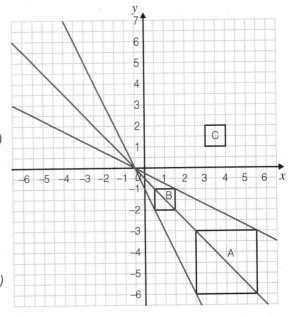

b) Describe the transformation which takes square A to B.

(2 marks)

23. Hon's horse jumps a 110cm high jump at an angle of 50°. The horse takes off 130cm away from the base of the jump.

By how many centimetres does the horse clear the jump?

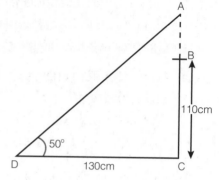

_____ cm *(2 marks)*

24. Here are some facts about the Earth:

Circumference at the Equator = 40075km

Radius = 6371km

The Earth spins on its own axis once per day. Dividing the circumference by 24 gives the speed at which the Earth spins around its own axis.

40075 ÷ 24 = 1670km per hour (rounded to the nearest whole number)

a) The Earth's orbit around the Sun is 940 million km. The Earth orbits the Sun once per year. In a normal year (not a leap year) there are 8760 hours.

Change each of these numbers to standard form and calculate the Earth's orbital speed in km per hour. Give your final answer in standard form.

Earth's orbital speed: _____ km/h *(2 marks)*

b) The volume of a sphere is $\frac{4}{3}\pi r^3$

Use the formula to calculate the volume of the Earth, taking $\pi = 3.14$ and converting the Earth's radius to standard form rounded to 2 significant figures. Give your final answer in standard form.

_____ km³ *(2 marks)*

1. a) 10:00 31st December 2019 ↺ 140/141
 b) 10 + 8 = 18 hours ↺ 6
 c) Christmas Island and Papeete/Fakaofo and Alofi/
 Funafuti and Baker Island (Time differences add up to
 24 hours) ↺ 6

2. a) $\frac{1}{3}$ ↺ 159
 b) $\frac{1}{4}$ ↺ 159
 c) $\frac{2}{6} = \frac{1}{3}$ ↺ 159

3. a) 27 ↺ 65
 b) 4 ↺ 66

4.

Length (cm)	Width (cm)	Area (cm²)	Perimeter (m)
5	3	**15**	**16**
6	4	24	**20**
6	**3**	18	18

(1 mark for each pair of correct answers) ↺ 144
↺ 14

5. No
$\frac{8}{12} = \frac{2}{3}$
$\frac{14}{42} = \frac{1}{3}$

6. a) 12% of 400 = 40 + 8 = 48 ↺ 24
 400 + 48 = 448kg
 b) 12% of 300 = 30 + 6 = 36 ↺ 25
 300 − 36 = 264kg
 c) 366kg = 122% ↺ 27
 366 ÷ 122 = 3
 3 = 1%
 So, plum trees produced 300kg of fruit last year.

7. a) Salad onions ↺ 188
 b) Brussels sprouts ↺ 188
 c) **70** per cent of the cucumbers eaten in the
 UK in 2011 were grown abroad. **(1 mark)**

8. a) 21 : 28 : 42 = 3 : 4 : 6 ↺ 28
 b) 156 ÷ 13 = 12 **(1 mark)** ↺ 28
 12 × 4 = 48 plain biscuits **(1 mark)**

9. a) This calculation tells us that each customer bought
 about **2 meals**. ↺ 42
 b) This calculation tells us that on average about ↺ 42
 18 customers per hour came to the takeaway.

10. a) $x = 20°$ ↺ 105/106 **(1 mark)**
 $DEB = 110°$ (alternate angles)
 $CAB = 70°$ (supplementary angles)
 $ACB = 20°$ (angles in a triangle) **(1 mark)**
 b) $2\frac{3}{4} \times 2$ ↺ 144/145

 $= \frac{11}{4} \times \frac{2}{1}$

 $= \frac{11}{2}$ **(1 mark)**

$\frac{11}{2} \div \frac{22}{7} = \frac{11}{2} \times \frac{7}{22}$

$= \frac{7}{4} = 1\frac{3}{4}$

AB = $1\frac{3}{4}$cm (or 1.75cm) **(1 mark)**

c) $4\frac{2}{3} \times 3\frac{1}{7}$ ↺ 144

$= \frac{14}{3} \times \frac{22}{7}$

$= \frac{44}{3}$

$= 14\frac{2}{3}$

So $ACDE = 14\frac{2}{3}$cm²

11.

Term 1	Term 2	Term 3	Term 4	Term n
7	9	11	13	$2n + 5$
13	11	9	7	**$15 - 2n$**
4	**11**	**18**	**25**	**$7n - 3$**
−4	**−11**	**−18**	**−25**	**$3 - 7n$**

(1 mark for each three correct answers) ↺ 79

12. a) (4, 4) and (5, 6) ↺ 83
 b) $2x - y = 8$ ↺ 84
13. a) 48 − 12 = 36 ↺ 198
 b) 620 **(1 mark)** ÷ 20 = 31 **(1 mark)** ↺ 198
14. a) 89.26 + 76.3 = 165.56 ↺ 17
 b) 84 − 28.739 = 55.261 ↺ 17
15. 4 1 3 9 **(1 mark)** and 4 3 1 9 **(1 mark)** ↺ 8–10
16. a) 5 × 49 = 245 ↺ 63
 b) 48 + 6 = 54 ↺ 63
17. ↺ 163

	Year 7	Year 8	Year 9	Total
WL	9	6	9	**24**
MP	10	7	7	**24**
LL	9	12	11	**32**
JT	2	3	5	**10**
Total	**30**	**28**	**32**	**90**

b) No ↺ 163
 In Year 7 the probability is $\frac{7}{28}$

 In Year 8 the probability is $\frac{7}{32}$

c) No ↺ 163
 $P(W) = \frac{24}{30}$

 $P(MP) = \frac{24}{30}$

 $P(W \text{ or } MP) = \frac{24}{30} + \frac{24}{30} = \frac{48}{90}$

 $\frac{48}{90}$ is about 50%, not 25%

18. $4x - 4y = 8$ ↻ 68

$x + 2y = 14$

$4x - 4y = 8$

$2x + 4y = 28$

Add the two equations:

$6x = 36$

$x = 6$ **(1 mark)**

$6 + 2y = 14$

$2y = 14 - 6$

$2y = 8$

$y = 4$ **(1 mark)**

19. a) Anuja's dad is correct.   ↻ 144/43/49

Perimeter of field $= 2(50 + 80) + (6 + 6)$

$= 260 + 12$

$= 272m$

$272 \div 6.5 = 41.846$

Because Anuja needs 0.846 of a roll, the answer must be rounded up to 42 rolls.

b) Anuja's dad is correct   ↻ 144/43/49

$272 - 4.2 = 267.8$

$267.8 \div 6.5 = 41.2$

Again this must be rounded up to 42 rolls.

c) Area $= (80 \times 50) - (20 \times 6)m^2$ ↻ 144/41/42/49

$= 4000 - 120$

$= 3880m^2$

Amount of fertiliser $= 3880 \times 0.2$

$= 776$

$776 \div 10 = 77.6$

This must be rounded up, so Anuja will require 78 bags.   **(2 marks)**

20.   ↻ 40–42

Ferry	$= £193.45 \times 2$	$= £386.90$
	$= £98.65 \times 2$	$= £197.30$
		$£150.00$
Total		$= £734.20$
Air	$= £230.50 \times 2$	$= £461.00$
	$= £75 \times 2$	$= £150.00$
		$£40.00$
		$£14.00$
Total		$= £665.00$ ✓

Family room	$= £300$
Total	$= £300$
Hotel rooms	$= £180$
	$£90$
	$£28$
Total	$= £298$ ✓

Family pass	$= £86 + (25\%$ of $£86)$
	$= £86 + £64.50$
Total	$= £150.50$ ✓
Daily tickets	$= (£27 \times 4) + (£16 \times 4)$
	$= £108 + £64$
Total	$= £172$

So,

Transport: Air   **(1 mark)**

Accommodation: Hotel rooms   **(1 mark)**

Entry costs: Family pass   **(1 mark)**

21. a) £25   ↻ 88

b) $10 \div 2 = 5$ ↻ 88

c) Charge $= 5x + £25$ ↻ 88

d) $(45 \times 5) + 25 = £250$ ↻ 88

22. a) Translation by a vector of $\begin{pmatrix} 5 \\ 6 \end{pmatrix}$ ↻ 125

b) Enlargement by a scale factor of $\frac{1}{3}$ around a centre of $(-1, 0)$.   ↻ 126

23. $\tan 50° = \dfrac{\text{opposite}}{\text{adjacent}}$ ↻ 114–116

$= \dfrac{x}{130} \quad \tan 50° \times 130 = x$

$x = 154.92$ (rounded to 155cm)   **(1 mark)**

$155 - 110 = 45cm$

Hon's horse clears the jump by 45cm   **(1 mark)**

24. a) 940 million $= 940\,000\,000 = 9.4 \times 10^8$ ↻ 12/13

8760 hours $= 8.76 \times 10^3$ **(1 mark for changing both numbers to standard form)**

$\dfrac{9.4 \times 10^8}{8.76 \times 10^3}$

$= 1.07 \times 10^{(8-3)}$

$= 1.07 \times 10^5$

Earth's orbital speed: 1.07×10^5 km/h   **(1 mark)**

b) $\dfrac{4}{3} \times 3.14 \times (6.4 \times 10^3)^3$ ↻ 12/13 **(1 mark)**

$= \dfrac{4}{3} \times 3.14 \times 262.144 \times 10^9$

$= 1097.5095 \times 10^9$

$= 1.098 \times 10^{12}$ km^3

So volume of Earth $= 1.098 \times 10^{12}$ km^3 **(1 mark)**

124

...ation of a shape to make a similar

56

...nt that two or more things are equal.

...ng 50

...a rough calculation which may involve ...more approximations.

...ession 56

...lgebraic statement containing symbols and ...sibly numbers.

...haustive events 158

...vents that account for all the possible ...utcomes.

Face 100
One of the flat surfaces of a 3-D shape.

Factor 7
A number that divides exactly into another number.

Factorising 60
Separating an expression into its factors.

Finite set 163
A given number of members of a set.

Formula 56
A mathematical expression that is used to solve problems.

Frequency 168
The number of times that an event has occurred.

Frequency polygon 191
Joins the midpoints of class intervals for grouped or continuous data.

Function 57
This is a relationship between two sets of values.

Gradient 83
The slope of a line in relation to the positive direction of the x-axis.

Highest common factor 8
The largest factor that two numbers have in common.

Horizontal 82
A line that goes straight across from left to right.

Hypotenuse 110
The longest side of a right-angled triangle.

Hypothesis 178
A statement that can be tested to see if it is true.

Identity function 81
This is when one value is mapped onto itself.

Improper fraction 14
This is when the numerator is bigger than the denominator.

Independent events 166
If two events have no effect on each other, they are said to be independent.

Index 11
The power to which a quantity is raised.

Inequality 70
A statement that two or more things are not equal.

Infinite set 163
This is an unlimited number of members of a set.

Integer 6
A positive or negative whole number.

Intercept 84
The point at which a graph cuts the y-axis.

Isosceles triangle 98
A triangle with two equal sides and two equal angles.

Linear 65
Consisting of a line or having one dimension; a linear graph is a straight-line graph; a linear expression is one such as $2x + 3$, which on a graph gives a straight line.

Line graph 192
A graph formed by joining points with straight lines.

Line segment 97
A straight line which has finite length.

Locus 130
(Plural loci) The locus of a point is the set of all possible positions that the point can occupy, subject to some given conditions or rule. Loci is the plural of locus.

Lowest common multiple 8
The lowest number that is a multiple of both numbers.

Lowest terms 18
A fraction is in its lowest terms when it cannot be cancelled down any further.

Mapping 81
A relationship between one group of numbers and another group of numbers.

Mean 196
The sum of all the values divided by the number of values used.

Median 201
The middle value when a set of numbers is put in order of size.

Mixed number *14*
A number that includes both whole numbers and fractions.

Mode *197*
The value that occurs most often.

Multiples *7*
The numbers in the multiplication tables, e.g. multiples of 5 are 5, 10, 15, 20, … etc. since 5 will divide exactly into these numbers.

Multiplier *24*
The same as a scale factor.

Mutually exclusive *159*
Events that cannot happen at the same time.

Net *100*
A flat shape that can be folded into a 3-D solid.

Null set *163*
The set that contains no members.

Numerator *14*
The top part of a fraction.

Obtuse angle *104*
An angle measuring between 90° and 180°.

Outcomes *158*
The possible results of a statistical experiment or other activity involving uncertainty.

Outlier *197*
A value that 'lies outside' (is much smaller or larger than) most of the other values in a set of data.

Parallel lines *97*
Lines that never meet; they are always the same distance apart.

Percentage *21*
A fraction with a denominator of 100.

Perimeter *144*
The distance around the outside edge of a shape.

Perpendicular bisector *99*
A line which cuts another line exactly in half and at right angles.

Perpendicular lines *97*
Two lines are perpendicular to each other if they meet at 90°.

Pictogram *187*
A charts that uses symbols: each symbol represents a certain number of items.

Plan *101*
The view of a 3-D shape when looked down on from above.

Polygon *99*
A plane figure that has three or more straight sides; a regular polygon has all sides and all angles equal.

Population *179*
A word that is used to describe a set, collection or group of objects that are being studied.

Primary data *178*
Data that you collect yourself.

Prime number *8*
A number that only has two factors: 1 and itself.

Prism *100, 148*
A three-dimensional shape having the same cross-section throughout its length.

Probability *158*
The chance or likelihood of something happening.

Product *41*
The result of two or more numbers multiplied together.

Proper fraction *14*
A fraction in which the denominator is greater than the numerator.

Quadrilateral *98*
A polygon with four sides.

Questionnaire *181*
A sheet with questions, used to collect data.

Radius *99*
The distance from the centre of a circle to the circumference.

Range *196*
The difference between the highest and lowest numbers in a set of data.

Ratio *28*
A comparison between two quantities, which are measured in the same units

Reciprocal *9*
The reciprocal of a number $\frac{a}{x}$ is $\frac{x}{a}$

Recurring decimal *17*
A decimal number whose decimal places carry on repeating forever in a set pattern.

Reflection *124*
A transformation in which an object appears to have been reflected in a mirror.

Reflex angle *104*
An angle measuring between 180° and 360°.

Relative frequency *169*
This is used as an estimate of a probability.